# Compounding Machines

*The Math and Psychology of Long-Duration Wealth*

Joseph Trek

Published by NobleTrex Press



For permissions and other inquiries, write to:
P.O. Box 3132, Framingham, MA 01701, USA

# Contents

# Introduction

The most dangerous instinct you possess is the one that kept your ancestors alive.

Imagine you are standing on the African savanna fifty thousand years ago. A rustle in the tall grass suggests a predator. Your brain, evolved for immediate survival, floods your body with cortisol. You don't pause to calculate the statistical probability that the rustle is merely wind; you run. The cost of a false positive (running from the wind) is a few burned calories. The cost of a false negative (ignoring a lion) is death. Your biology is hardwired to react instantly to threats, to prioritize the short term over the long term, and to equate doing *something*-anything-with safety.

Now, fast forward to a Tuesday morning in the twenty-first century. Your smartphone buzzes. A news alert flashes red: "Markets Tumble as Inflation Fears Mount."

That same ancient circuit fires. Your cortisol spikes. The rustle in the grass is now a dip in the S&P 500, and your instinct screams at you to *do something*. Sell. Buy. Hedge. Tinker. Fix it.

But here lies the cruel irony of modern wealth: In almost every area of human endeavor-athletics, career, relationships-effort correlates with results. If you work harder, you generally get better outcomes. Investing is the singular exception. In the physics of long-duration wealth, activity is often the enemy of progress. The more you touch your portfolio, the more likely you are to break it.

This book is about building a system that allows you to defy your biology. It is about constructing a **Compounding Machine**-a rigorous mathematical framework designed to harness the exponential power of time-and then installing a behavioral firewall to protect that machine from its greatest threat: the person looking back at you in the mirror.

## The Invisible architecture of Wealth

Most finance books treat investing as a shopping trip. They obsess over the merchandise. They argue about whether you should buy Tech Stock A or Crypto Coin B, whether real estate is better than bonds, or which mutual fund manager has the hot hand. They focus on the "what."

This approach is fundamentally broken because it ignores the "how" and the "why."

If you gave a Formula 1 car to a driver who didn't understand how to shift gears or handle a corner, they wouldn't win the Grand Prix; they would crash into the first wall. Similarly, handing a high-performance portfolio to an investor without the requisite mental models is financial negligence. The assets (the car) matter, but the architecture of the machine and the temperament of the operator matter infinitely more.

We are going to take a different approach. We are going to treat wealth creation not as a series of lucky guesses, but as an engineering problem.

When you strip away the cable news noise and the Wall Street marketing jargon, successful investing dissolves into physics and psychology. The physics involves the mathematical laws of compounding, the relationship between risk and return, and the corrosive friction of fees and taxes. These laws are immutable. They work the same way for a teacher saving for retirement as they do for a sovereign wealth fund.

The psychology, however, is the variable. It is the chaotic element. It is the reason why the average investment fund returns 10% over a decade, but the average investor in that fund earns only 6%. That "behavioral gap" is the price of human emotion trying to navigate a mathematical environment.

This book is your blueprint for closing that gap.

## The Tyranny of the Present

Why is this so difficult? Why do smart, successful people-doctors, engineers, entrepreneurs-routinely incinerate their wealth in the stock market?

The answer lies in our perception of time. Humans are linear thinkers living in an exponential world. If you take thirty linear steps, you end up thirty meters away. If you take thirty exponential steps (doubling your stride each time), you don't end up across the street; you end up circling the Earth twenty-six times.

Our brains cannot intuitively grasp that magnitude. We understand "saving for a rainy day," but we struggle to visceralize how a dollar saved today, compounded at 7% for forty years, transforms into fifteen dollars of purchasing power. Because we cannot *feel* the math, we discount it. We prioritize the immediate gratification of spending or the immediate relief of selling during a downturn.

We are also fighting a war against an invisible enemy: entropy. In the financial world, entropy takes the form of inflation. It is the silent heat death of your money. If you bury your cash in the backyard to keep it "safe," you are guaranteeing a loss. A 3% inflation rate cuts the value of your money in half every twenty-four years. Doing nothing is not a neutral act; it is a decision to slowly go broke.

Therefore, you have no choice but to participate. You must build a machine that outpaces the erosion of your purchasing power. You must expose your capital to risk to preserve its value. This paradox-that safety is dangerous, and risk is the only path to safety-is the first mental hurdle we must clear.

## The Blueprint

We will not be relying on "hot tips" or forecasting the future. Predicting the future is impossible, and believing you can do it is a sign of amateurism. Instead, we will rely on probability, structure, and patience.

Our investigation begins with **The Physics**. Before we buy a single share, we must understand the terrain. We will strip down the concept of "returns" to its chassis. You will learn why "average" returns are a mathematical fiction that rarely happen in any given year, and why the sequence of those returns can make or break a financial plan. We will look at the measuring stick itself-money-and realize that tracking your wealth in nominal dollars is like measuring your height with a ruler that shrinks every year.

Once we understand the laws of motion, we will examine **The Components**. Every asset class-equities, bonds, real estate-has a specific job. We will move beyond the superficial labels and look at the engine mechanics. We won't just say "stocks go up over time"; we will dissect *why* they go up, analyzing the cash flows, dividends, and earnings growth that drive value. We will look at debt not as a terrifying burden, but as a misunderstood tool for stability and yield. We will debunk the myths of gold and crypto, and look at what actually protects you from inflation.

With the components laid out, we will turn to **Architecture**. This is where art meets science. How do you combine volatile assets to create a stable portfolio? This is the magic of diversification-the only "free lunch" in finance. We will explore the efficient frontier and the mathematics of correlation, proving how adding a risky asset to a risky portfolio can, counterintuitively, lower the overall risk. We will also discuss the dangerous allure of leverage, which can supercharge your machine or blow it apart completely.

## The Ghost in the Machine

If the first half of this book is about the hardware, the second half is about the software: you.

You can have the most elegant, mathematically optimized portfolio in the world, framed on your wall and backtested to perfection. But if you panic and sell everything when the market drops 20%, that portfolio is worthless. **The Behavioral Firewall** is your defense against self-sabotage.

We will explore the strange wiring of the human mind. You will learn about "loss aversion," the phenomenon where the pain of losing $1,000 is psychologically twice as intense as the pleasure of gaining $1,000. We will examine "recency bias," which convinces us that whatever happened yesterday will continue happening forever. We will dissect the ego and the dangerous illusion of skill that leads to overconfidence.

Most importantly, we will move from diagnosis to prescription. It is not enough to know you are biased; you need systems to counteract those biases. We will design "guardrails"-rules and precommitments that tie your hands when you are tempted to meddle. We will automate your decisions so that your wealth accumulation happens in the background, requiring zero willpower to maintain.

## Survival of the Fittest

In the long run, survival is the only stat that matters. The history of finance is littered with the corpses of brilliant strategies that couldn't survive a bad year.

We will look at the **Friction** that slows your machine down. Wall Street is an industry designed to transfer wealth from your pocket to theirs via fees, spreads, and commissions. We will learn how to minimize these costs relentlessly. We will also tackle the government's share-taxes-and how to structure your assets to keep more of what you earn.

Finally, we will integrate this into the **Lifecycle of Money**. The way you manage your machine at age thirty should be radically different from how you manage it at age seventy. We will discuss the "Glide Path" of de-risking as you age, and the terrifying "Sequence of Returns Risk" that keeps retirees awake at night. We will solve the puzzle of how to turn a pile of assets into a reliable paycheck that you cannot outlive.

## A Note on "Boring"

If you are looking for excitement, go to a casino. If you are looking for social validation, join a country club. If you are looking to build multi-generational wealth, prepare to be bored.

Real investing-the kind that compounds over decades-is roughly as exciting as watching paint dry or grass grow. It is a process of aggressive patience. It requires you to be contrarian when the world is euphoric and stoic when the world is panicking.

The media industry is built to sell ads, and "Everything is Fine, Keep Saving" is a terrible headline. They need you to be anxious. They need you to feel like you are missing out. They need you to treat your portfolio like a video game.

This book is permission to opt out of that game. It is an invitation to play a different game entirely-a single-player game where the only opponent is your own impulse control.

By the end of this book, you will not have a stock tip that will double your money next week. You will have something far more valuable: a mental framework that renders you immune to the noise. You will understand that volatility is the price of admission for higher returns, not a sign that something is broken. You will view a market crash not as a disaster, but as a sale on future cash flows.

You are about to build a machine that works while you sleep, while you work, and while you live your life. It is a machine made of math, discipline, and time.

Let's look under the hood.

Chapter 1

# The Physics of Long-Duration Wealth

*We often treat investing like an art form-subjective, emotional, and prone to trends. But long-duration wealth behaves more like physics. It is governed by invisible laws: gravity (inflation), momentum (compounding), and friction (taxes and fees). If you fight these laws, you will lose. If you understand them, you can build a machine that works for you even while you sleep. This chapter hands you the schematics.*

## 1.1 Defining the Destination and the Timeline

On paper, the sentence looks responsible: "I want to be financially independent someday." In real life, it behaves like fog. Fog doesn't tell you whether to bring a jacket, whether to leave early, or whether the road is icy. A portfolio can't optimize for "someday" any more than an engineer can design a bridge for "some cars, sometimes." The market doesn't reward vague bravery; it rewards matching the job to the tool.

The first job of a long-duration investor is not picking stocks. It's choosing *when money must work* versus *when money must be available*. Those are different species of dollars. Treat them as interchangeable and you end up with the classic tragedy: selling a long-term compounding engine at the exact moment it's temporarily ugly, because a short-term bill showed up with a due date and no patience.

Think about two people with the same net worth and the same "risk tolerance" (whatever that means on a questionnaire). One is 29 with a stable job, no debt, and a paid-off car. The other is 42, self-employed with volatile income, a mortgage, and a child who will start college in six years. Same account balance; wildly different *risk capacity*. Risk tolerance is how much volatility you think you can stomach. Risk capacity is how much volatility your life can absorb without forcing you to do something financially destructive. The market mostly punishes people who confuse the two.

A useful way to get out of the fog is to stop calling everything "savings" and start naming the liabilities. A *liability* here doesn't mean debt. It means an obligation your future self will experience as non-negotiable: rent, a roof repair, a tuition bill, the down payment you swore you'd have by next spring. The word matters because liabilities have deadlines, and deadlines turn market volatility from an academic concept into a calendar problem. A liability due in 14 months is not "long-term" just because you *feel* long-term about it.

Now the engineering move: you don't invest for "the future." You invest for *a stack of futures*. Each future has a date, a probability of happening, and a penalty for being underfunded. Your emergency fund is a future. Replacing your car is a future. A sabbatical, a home purchase, a parent's care, a retirement income floor—all futures. The moment you accept that you are financing a schedule (not a dream), your portfolio stops being a personality test and becomes a system.

The reason this matters so much is simple and quietly brutal: markets are not obligated to cooperate on your timeline. Stocks can be down for years. Bonds can disappoint. Cash can erode. None of this is scandalous; it's the normal weather of capitalism. The scandal is building a plan where you *must* sell risky assets on a specific date to pay a bill. That's not investing; that's borrowing volatility.

Start by carving your financial life into buckets based on *when* the money is needed, not what the money is "for." The labels can be plain:

- *Safety money*: dollars you may need soon and cannot afford to have missing when the need arrives.
- *Stability money*: dollars you probably won't need this year, but you can imagine needing within a handful of years.
- *Growth money*: dollars you are deliberately giving time to compound, because time is the whole point.

The bucket names are less important than the logic: different timelines demand different behavior, and behavior is the hidden variable in every return chart.

A common mistake is to make the buckets about emotions: "This is my conservative account" versus "this is my aggressive account." That framing fails the first time life gets inconvenient. Make them about dates. Dates don't negotiate. Dates also prevent you from doing something psychologically seductive but mathematically poisonous: treating a long-term portfolio like an ATM during a short-term pinch.

Consider the down payment trap. A couple wants a house in three years. They put the down payment in a stock index fund because the market has been generous lately and cash yields feel insulting. Two years later, the market is down 30% and mortgage rates have changed. They can either delay the house (life plan rupture) or sell at a loss (portfolio rupture). The problem wasn't bad luck; it was a mismatch between the asset and the deadline. A three-year horizon is not "long-term" for stocks, no matter how disciplined you feel in January.

Or consider the opposite trap: someone in their 20s keeps everything in cash because they "don't like risk." Ten years pass. They did avoid volatility—and quietly paid for it by giving up the one resource they cannot buy back later: time. They didn't just miss returns. They missed *compounding runway*. Cash is a wonderful tool for short-term certainty. It's a leaky container for long-term purchasing power. When a person uses it as a long-term plan, they are choosing safety in the present at the expense of fragility in the future.

The timeline isn't only about goals. It's also about *liquidity*: how quickly you can turn an asset into spendable dollars without taking a painful haircut. Liquidity is the difference between solving a problem with a transfer and solving it with a regret. Cash is liquid. A brokerage account is usually pretty liquid, but the price you get depends on the market that day. A house is illiquid. A private business interest can be very illiquid. Even within a liquid brokerage account, liquidity isn't just "can I sell?" It's "can I sell *without breaking the plan*?"

That's why an emergency fund is more than a personal finance cliché. It's an *anti-forced-seller device*. If your only source of liquidity is your growth portfolio, every job loss, medical bill, or surprise move turns into a high-stakes market timing decision. In a bull market you'll feel clever. In a bear market you'll feel cursed. The real issue is structural: you built a system that requires good luck.

Risk capacity is the bridge between your timeline and your asset choices. People often frame capacity as a number: "I can handle a 20% drop." But capacity is more like a set of shock absorbers:

- How predictable is your income?
- How flexible is your spending?
- How large are your near-term obligations?
- How quickly could you raise cash without selling investments?

The same market drawdown is a paper cut to one household and a career-altering wound to another. The market move is identical; the household balance sheet is not.

A clean way to make this concrete is to write down your "cannot-fail" expenses for the next 12 months. Not "what you spend when life is good," but what you must pay to keep your life stable: housing, food, insurance premiums, basic transportation, minimum debt payments, essential healthcare. Then ask a slightly uncomfortable question: *If my income were disrupted for six months, where would the money come from?* If the answer is "I'd sell investments," then your portfolio is not yet an engine. It's a hostage.

Once you've protected the near-term, the long-term becomes easier—not because it's guaranteed, but because you can afford to stay rational when it's temporarily ugly. This is the quiet superpower of matching assets to liabilities: it converts volatility from an emergency into information. When your growth money is genuinely long-duration money, a bear market becomes a clearance sale, not a threat to your rent.

There's also a psychological reason timelines matter: humans are wired to confuse motion with progress. A portfolio that moves around feels alive. A savings account feels boring. Most bad investing is a search for the emotional payoff of activity: the feeling that you're doing something sophisticated. Timelines short-circuit this. When you know which dollars have deadlines, you stop trying to entertain yourself with them. You treat them like oxygen tanks. You don't day-trade oxygen.

The destination has to be defined with the same ruthless clarity. "Retire early" can mean at least three different things:

- Stop working entirely.
- Stop needing a high salary (work becomes optional).
- Keep working, but with bargaining power (you can walk away from a bad situation).

All three are valid. They just require different amounts of money and different degrees of liquidity. A person who wants optionality needs a different structure than someone who wants complete detachment from wages. Without that clarity, you can end up over-saving (living smaller than necessary) or under-saving (building a beautiful spreadsheet that collapses under real bills).

Engineers don't just specify the destination; they specify the tolerances. In investing, tolerances are rules that prevent a bad week from becoming a bad decade. A simple tolerance is: "I will not invest money in stocks if I may need it within five years." Another is: "My emergency fund will cover X months of cannot-fail expenses." These are not universal commandments; they are guardrails that turn your plan into something you can follow when you are tired, stressed, or tempted.

Writing those guardrails down changes your behavior in a way that will feel almost embarrassingly practical. It removes the need to renegotiate with yourself. Most financial mistakes are not made because people can't do math. They're made because people are forced to decide under pressure. A written rule turns a pressured decision into a routine action.

One more subtlety: timelines are not fixed. Life has a habit of moving the goalposts. That's not failure; that's biology. People get married, divorced, sick, inspired, burnt out, promoted, laid off, relocated. The plan has to be built to survive timeline drift. This is another reason to separate buckets by horizon: you can adjust the schedule without tearing apart the engine. If you keep everything in one undifferentiated pile, every life change becomes a portfolio crisis.

A concrete example: imagine you're 35 and want to "maybe" buy a home. If you don't know whether "maybe" is two years or ten years, you can't choose an asset mix with integrity. The fix isn't predicting the future; it's building a two-part structure. Put the portion that would become a down payment on a short horizon into safety/stability. Keep the rest in growth. If the home plan becomes real, you have money that won't be hostage to the market. If the plan fades, you haven't strangled the long-term engine with excessive conservatism.

Another example: college funding. A child's tuition schedule is the rare life event that arrives like a metronome. If you plan to pay meaningful tuition in years 13 through 16, the timeline is baked in. That doesn't mean "no stocks ever." It means the closer you get to the tuition years, the more you should shift those specific dollars toward assets with less chance of being down right when the bill arrives. This isn't about being timid; it's about not building a plan where your kid's freshman year depends on the S&P 500's mood in August.

If this starts to sound like a lot of moving parts, good. Real life has moving parts. But you can keep the system simple by using a single question to categorize each dollar: *When could I need this money, even if life misbehaves?* Not "when do I plan to need it" (plans are optimistic), but when you could need it if the world applies pressure. That question naturally pushes emergency funds and near-term obligations into safer assets, and frees the rest to do what it's meant to do: compound without being interrupted.

One last mental model: imagine your financial life as having two jobs. Job one is *staying in the game.* Job two is *winning the game.* Staying in the game is about liquidity, buffers, and deadlines. Winning is about compounding over decades. Most people flip the order. They try to win first, reach for returns, and then discover they can't stay invested when the first real storm arrives. The pros build the bunker first, not because they're pessimists, but because they want the freedom to take intelligent risk with the money that can afford it.

*Actionable takeaway:* Open a document and create a one-page "timeline map." List every foreseeable cash need in the next 0–2 years, 3–7 years, and 10+ years, with a rough dollar amount next to each. Then circle the items that would be disastrous to miss (rent, insurance, taxes, essential repairs). Commit to funding those circled items with money that prioritizes availability over return. Whatever remains in the 10+ year bucket is your true growth capital—money you can invest aggressively enough to benefit from compounding, because you've removed the most common reason people sell at the worst possible time.

## 1.2 The Engine: Compounding and Time Value

A strange thing happens when you first start investing: you do everything "right" and nothing impressive happens. You contribute. You rebalance. You ignore headlines. You watch your account wobble like a grocery cart with one bad wheel. Then, one day years later, you open the statement and realize the portfolio earned more in a year than you did by grinding out deposits. It doesn't feel gradual. It feels like you crossed an invisible border.

That delayed payoff isn't a motivational poster. It's the mathematics of multiplication. Compounding is what you get when gains become capital, and capital becomes a larger base for the next gain. The engine is simple enough to explain to a middle-schooler and sneaky enough to fool a grown adult with a mortgage: it rewards patience in a way that looks like nothing for a long time, then looks like luck.

If you want the cleanest version of the engine, start with one dollar. Let it grow at a rate $r$ per year. After one year you have $1(1+r)$. After two years you have $1(1+r)^2$. After $t$ years you have

$$(1+r)^t.$$

That exponent—that innocent little $t$—is why time is not just an input. It's the amplifier.

Most people, even very smart people, are linear thinkers. Linear thinking is what your brain uses to survive: if you drive 60 miles per hour, you go 60 miles in an hour. If you save $500 a month, you save $6,000 a year. Linear thinking is great for carpentry, cooking, and commuting. It is terrible for understanding why a 30-year investor and a 10-year investor can make the same monthly contribution and end up living in different financial universes.

To see the difference, compare two worlds that look similar early on. In world one, you save cash under a mattress (or in an account that merely holds value). Your wealth grows in a straight line: contributions add, and that's that. In world two, you invest and earn a return that compounds. Early on, they look like twins. Later, they look like a magic trick.

Make the numbers concrete. Suppose you invest \$500 per month, or \$6,000 per year, for 30 years. Assume a 7% annual return (a round number that makes the point without pretending to predict the future). If you held the cash with no return, you'd have \$180,000 after 30 years. That's not nothing; it's a respectable pile of discipline.

Now put the same schedule into a compounding engine. The future value of a stream of equal annual contributions $C$, invested at rate $r$ for $n$ years, is

$$FV = C \cdot \frac{(1+r)^n - 1}{r}.$$

With $C = \$6{,}000$, $r = 0.07$, and $n = 30$, you land around \$566,000. Same worker. Same savings habit. A very different ending.

The part that surprises people isn't just the final number. It's where the number comes from. In the early years, the portfolio is mostly you—your deposits, your restraint, your choosing "later" over "now." Then comes the shift. The portfolio becomes mostly the portfolio. Not because you got smarter or the market got kinder, but because the base grew large enough that a normal percentage move produces real money.

This shift has a name in polite investing conversations: the "inflection point." In private, it's the moment compounding starts to feel like cheating.

Here's the trick your brain plays on you: you think compounding is about getting rich quickly. It's almost the opposite. Compounding is about getting rich *slowly*, then all at once. That "all at once" is an illusion created by exponential curves. The growth is smooth in math and lumpy in emotion.

A useful frame is to stop thinking about compounding as "interest on interest" and start thinking about it as *a machine that changes what matters*. In the beginning, what matters is the contribution rate. In the middle, what matters is staying invested. In the later years, what matters is avoiding catastrophic mistakes. The engine does not forgive you equally at every stage. A bad decision in year 3 is annoying. A bad decision in year 28 can be life-altering.

The difficulty is that the early stage feels like a scam. If you invest $6,000 a year and earn 7%, the first year's gain is only a few hundred dollars. You can't "feel" a few hundred dollars, not in a world where a single car repair can eat it. You start to wonder if the whole thing is just an elaborate way to keep you from enjoying your money.

But the math is doing what it always does: it's charging you an entry fee called time. The initial years are not supposed to look glamorous. They are supposed to build the base. A compounding engine without a base is like a jet engine without a runway: impressive in theory, useless in practice.

That's why so many investing stories sound like overnight success and are, on inspection, 20-year success with selective memory. People remember the moment the account balance became meaningful. They forget the decade when it felt like pushing a boulder that didn't want to move.

One reason the engine feels so weird is that returns scale with the portfolio, while contributions usually scale with your paycheck. Your paycheck tends to rise gradually. Your portfolio, if it compounds, rises multiplicatively. Eventually those two curves cross. When they cross, your financial identity changes. You stop being someone who is building wealth primarily through labor and start being someone whose capital is doing meaningful work.

That crossover moment is so important that it's worth drawing.

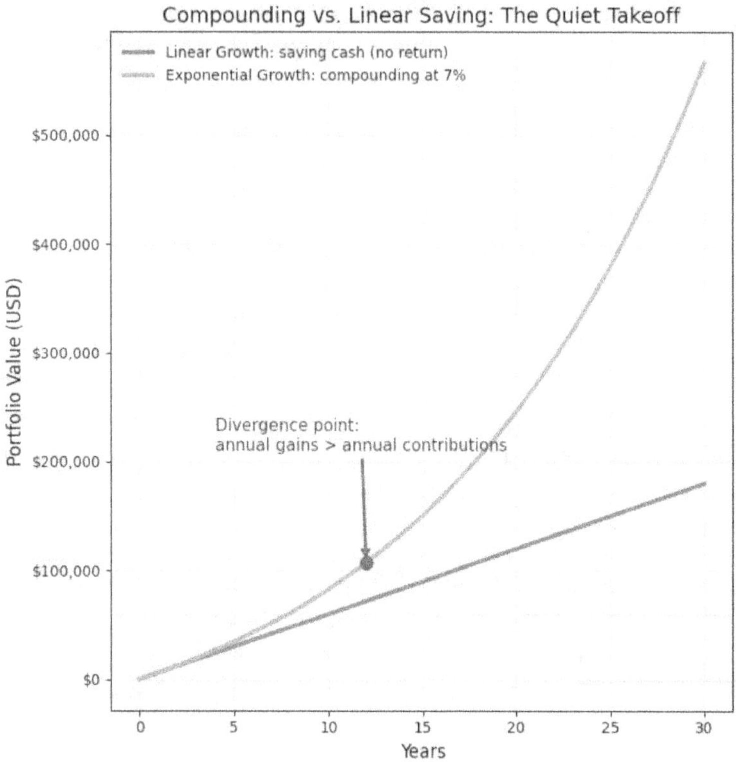


That red dot is the moment many people misunderstand retroactively. They look back and say, "The market finally started working." The market was always doing what markets do. What changed was the scale of the base. At small balances, returns are financially polite. At larger balances, the same percentage becomes a loud number.

Notice something else: the linear line never changes character. It is honest and predictable. The compounding curve changes character dramatically even though the return assumption stayed constant. This is why investors who are new to the engine are so vulnerable to impatience. They are evaluating an exponential machine using linear instincts.

There's a second misunderstanding hiding in the word "time value." People hear it and think it's finance-speak for "waiting." It's more precise than that. Time value is the idea that a dollar today is not the same as a dollar later because today's dollar can be deployed. It can earn. It can compound. It can also protect you from having to borrow at awful rates or liquidate assets at awful moments. Time gives money *options*. Options are value.

This is why delaying investing is costly in a way that doesn't feel costly. If you skip investing for five years, you don't just miss five years of returns. You miss the *final years' returns on those early contributions*. The loss shows up decades later, wearing the disguise of "I guess I didn't make as much as I hoped."

A classic illustration is two savers:

- Saver A invests $6,000 a year from age 25 to 35, then stops contributing entirely and leaves it invested.
- Saver B waits and invests $6,000 a year from age 35 to 65.

Saver B contributes three times as long. Yet Saver A often ends with a surprisingly competitive result, depending on the return environment, because Saver A bought the scarce resource: early years.

The moral isn't that contributions don't matter. The moral is that *early contributions are disproportionately powerful*. They are the seeds that get multiplied the most times.

There's a mental shortcut called the Rule of 72 that makes the time value feel less abstract. Divide 72 by your annual return percentage to estimate how long it takes to double. At 8%, money doubles roughly every 9 years ($72/8 = 9$). At 6%, it's about 12 years. It's not exact, but it's good enough to change how you feel about time. If your long-duration money can double multiple times, you stop treating decades as blank space and start treating them as a production schedule.

This is also where people get seduced into a dangerous game: chasing a higher return because they want a shorter doubling time. They see that 12% doubles in about 6 years and start shopping for 12% like it's a price tag. The engine's cruel joke is that higher expected return usually arrives packaged with higher volatility—and volatility is the ingredient that can shake you out of the machine. A theoretical 12% you can't stick with is less valuable than a 7% you can hold through boring years and bad headlines.

Compounding is not just math; it is also behavior. Your portfolio compounds only if *you* compound—if you keep adding fuel, keep your hands off the emergency lever, and avoid the two classic self-inflicted engine failures: panic selling and performance chasing. The mathematics assumes you stay in the game. Real life tries to interrupt you with fear, envy, and the intoxicating feeling that you can outsmart randomness.

Here's a sobering detail that turns "time in the market" from a slogan into an engineering constraint: missing a handful of the market's best days can damage long-term results significantly. The best days often cluster around the worst days, because markets snap back violently when fear peaks. If you flee during the crash, you often miss the rebound. The person who "waits for things to feel safer" is trying to buy certainty with time, and paying for it with return.

The point is not that you should never change your allocation. The point is that you should not make allocation decisions in response to your nervous system. The engine can handle turbulence. It cannot handle you yanking out parts mid-flight.

A long-duration investor needs a different definition of progress. Progress is not "my account is up this month." Progress is "I increased the base," "I lowered the friction," "I made it more likely I will stay invested." Those are engine-building actions. They are not exciting at parties. They are exciting at 65.

Compounding also creates a weird emotional mismatch: the later years matter more, but the earlier years require more faith. In year

1, a 20% drop might mean you "lost" $800. In year 25, the same percentage drop might mean a six-figure drawdown. The later-stage investor faces bigger absolute swings. That can feel like higher risk, even if it's the same percentage risk. The right response isn't to avoid growth assets entirely; it's to make sure the dollars with deadlines aren't riding in the same vehicle as the dollars meant to compound. When the timelines are clean, you can let the engine be an engine.

If you want one sentence that captures the physics: compounding turns *time* into *leverage*. Not the borrowing kind—the quiet kind. The kind where a small, consistent force applied over a long distance moves something enormous. It's not dramatic in the moment. It's dramatic in the mirror.

*Actionable takeaway:* Pick one "growth bucket" account that you genuinely do not need for at least 10 years, and write a one-line operating rule you can live with: "I will contribute $___ per month automatically, and I will not change the investment mix based on market news." Then run the Rule of 72 on your assumed long-run return to estimate how many doublings that bucket could reasonably experience over your timeline. If the number of doublings is less than two, the lever isn't broken—the timeline is. Adjust the timeline, the contribution, or the expectations, but don't ignore the math.

## 1.3 The Measuring Stick: Real vs. Nominal Progress

A friend texts you a screenshot: their portfolio "hit $500,000." Confetti emoji. Victory lap. The number is clean, round, and emotionally loud. The only problem is that a portfolio balance is not wealth in the way your life experiences wealth. Your life doesn't eat dollars; it eats groceries. It doesn't vacation in account statements; it vacations in airline seats and hotel nights. If the price of everything you want doubles while your account doubles, you didn't get richer. You stayed in place and threw a party.

That's the first mental shift: the measuring stick is not dollars. The measuring stick is *purchasing power*—what your money can buy in the world you actually live in. "Nominal" is the word for raw dollars, unadjusted. "Real" is the word for dollars translated into what they can buy after inflation. Nominal numbers are the scoreboard. Real numbers are the game.

Inflation is often treated like a villain that appears only in dramatic years. In practice, it's more like gravity. You can ignore it for a while, but it keeps working. It's the silent force that makes "the same salary" feel smaller, makes your favorite cheap meal quietly become a treat, and makes a retirement target that sounded enormous at 30 sound merely adequate at 60.

A useful way to think about inflation is as a guaranteed negative return on cash *in real terms*. Cash can still be a great tool for stability and timing—you need it to avoid becoming a forced seller—but if you leave long-duration money sitting in it, inflation will slowly, reliably strip it for parts. Not always quickly enough to feel like danger. Quickly enough to matter.

The great confusion is that nominal progress is seductive. It's measurable, it's easy, and it's always happening. Even in a mediocre decade, wages tend to rise, rents tend to rise, home prices tend to rise, and account balances may rise. The numbers keep moving, so it feels like forward motion. Real progress is harder. It requires translating every number into a common unit: today's dollars.

Here's the uncomfortable truth: most long-term goals people care about are already real goals, even if they talk about them in nominal terms. "I want $80,000 a year in retirement" is not really about the number 80,000. It's about a lifestyle: housing, healthcare, travel, generosity, safety. The lifestyle is the goal. The number is a placeholder that must be inflation-adjusted to remain meaningful.

So how do you actually adjust your thinking without turning your financial life into a graduate seminar? You pick a reference year and you keep coming back to it. "Today's dollars" is a powerful anchor.

When you hear that inflation is 3% or 4%, you don't need to predict the exact future. You need to respect the direction and the magnitude.

One reason inflation deserves this respect is that it compounds too. Compounding isn't morally aligned with investors. It works for whatever rate you feed it. A 3% annual inflation rate doesn't sound like a monster. Over 20 years it cuts purchasing power roughly in half. The math is the same engine as investing, pointed the other way.

The simplest formula is:

$$\text{Real value in today's dollars} = \frac{\text{Nominal dollars}}{(1+i)^t},$$

where $i$ is the inflation rate and $t$ is years. If you want to know what $100,000 will feel like after 20 years of 3% inflation:

$$\frac{100{,}000}{(1.03)^{20}} \approx 55{,}000.$$

Same nominal number on the screen; much smaller life behind it.

This matters because the compounding engine you built in your growth bucket is not trying to beat cash. It's trying to beat inflation *after* fees and taxes, because that's the hurdle rate for real wealth. If your portfolio earns 6% nominal and inflation is 3%, your real return is roughly 3% before considering taxes and costs. That "roughly" is doing a lot of work, but it's accurate enough for a mental model: real return is what's left after the world takes its share.

Notice what this does to the idea of being "conservative." A portfolio that avoids volatility by hugging cash can look conservative in nominal terms while being quietly aggressive in real terms—aggressive in the sense that it is making a strong bet that inflation won't matter. That bet is rarely stated, rarely intentional, and often expensive.

Inflation also changes how you interpret market returns. If the stock market is up 10% in a year with 8% inflation, that is not a 10% victory in real life. It's closer to a 2% gain in purchasing power, before taxes. Conversely, a flat nominal year during low inflation can be less damaging than it feels. Your brain reacts to the nominal chart. Your future lifestyle reacts to the real chart.

This is why investors sometimes feel like they are doing fine for years and then feel abruptly behind. They were tracking the wrong metric. They watched the nominal balance rise and assumed it represented progress. In reality, they were running on a treadmill where the speed slowly increased. The sweat was real; the distance wasn't.

There's a second twist: inflation is not a single, universal experience. "CPI"—the Consumer Price Index—is the most common public yardstick for inflation in the United States, built from a large sample of consumer prices. But your personal inflation rate can differ meaningfully based on where you live and what you spend on. A retiree who spends heavily on healthcare can feel a different inflation reality than a young renter whose biggest line item is rent. CPI is useful as a shared language. It's not a perfect mirror of any one household.

That imperfection can make some investors throw up their hands: "If it's not exact, why bother?" Because precision isn't the point. Calibration is. If you aim without accounting for inflation, you are not being simpler; you are being systematically wrong in one direction. Even a rough adjustment is better than none, the same way a map with a small scale error is still better than walking with your eyes closed.

Once you accept real-vs-nominal, you unlock a cleaner way to plan: express goals in today's dollars, and let time do the translation. If you want the equivalent of $80,000 per year of spending power, keep it as $80,000 in your plan. Then when you project forward, you either:

- use a *real return* assumption (returns after inflation), or

- use nominal returns and nominal spending with an explicit inflation rate.

Either method can work. The mistake is mixing them—projecting nominal portfolio growth and comparing it to a spending number that is secretly in today's dollars.

This is also where the psychological advantage of "today's dollars" shows up. Nominal retirement targets get absurd quickly. If you tell a 30-year-old they might need $3 million to retire, they might feel either doomed or reckless. If you tell them they need enough to fund $80,000 of spending power in today's dollars, that's intelligible. It maps to a lifestyle. It invites planning rather than panic.

There is an emotional trap hiding in inflation math: it can make the future feel hopelessly expensive. You do the projection and see that prices "triple" over a lifetime and it feels like you're being asked to outrun a moving walkway that never stops. The counterweight is remembering that your income is not static either. Your earnings, your skills, your ability to save, and your portfolio contributions typically rise over time. Inflation is a headwind, yes. But you are not a stationary object. The goal is not to defeat inflation with heroics. It's to build a system that naturally adapts.

A practical way to make inflation real without obsessing is to name your "minimum acceptable real return" for long-duration money. For many investors, the job of the growth bucket is to beat inflation by enough to grow purchasing power meaningfully over decades. That surplus—real return—is what funds better choices later: earlier retirement, more generous giving, more resilience to healthcare surprises, a home that doesn't require a commute you hate.

Framing it this way also clarifies why fees, taxes, and bad behavior are so deadly. If inflation is 3% and your portfolio earns 6% nominal, you might think you have a 6% engine. You don't. You have a 3% real engine before frictions. Add a 1% all-in cost structure and some avoidable taxes and a few panicked decisions, and you can turn a meaningful real compounding rate into something barely above zero. Nominally, the portfolio can still look like it's "growing." Realistically, you're pedaling hard and staying put.

Inflation also changes how you evaluate safety. People often treat government bonds, high-quality bonds, and cash as "safe" because their nominal value is stable. But stability of *nominal principal* is not the same as stability of *future spending power*. A portfolio meant to fund a liability next year needs nominal stability. A portfolio meant to fund groceries 30 years from now needs real stability. Those are different tasks.

This difference is why long-duration planning tends to produce a counterintuitive conclusion: for far-future goals, avoiding all volatility can be a form of risk. It's the risk of quietly failing the real objective. The trick isn't to worship volatility; it's to accept the kind of risk that can plausibly be compensated over time, while insulating yourself from the kinds of risk that force bad timing.

How do you put this into a habit, not just a concept? You make your tracking real. If you're measuring your portfolio monthly, you're already tempted to confuse noise with signal. Don't add the inflation illusion on top of that. Pick one day a year—your birthday, tax day, New Year's—and write down:

- your nominal net worth,
- an inflation-adjusted estimate of your net worth in today's dollars, and
- your goal numbers in today's dollars.

The annual cadence matters. Inflation is slow. Compounding is slow at first. Measuring too frequently trains you to feel anxious about things that require patience.

If you want a quick, clean tool: assume a long-run inflation rate for planning purposes (many people use 2%–3% as a ballpark) and keep it consistent. Consistency is not about being right; it's about avoiding self-deception. When your plan looks too hard, the temptation will be to assume lower inflation. When your plan looks too easy, the temptation will be to ignore inflation altogether. A steady assumption keeps you honest and forces the real levers to show themselves: savings rate, timeline, asset allocation, and behavior.

Inflation is the reason "I'm a millionaire" is not a permanent identity. In 1960, a million dollars bought a small empire. In 2026, it can buy a house in many cities, a decent retirement in some, and a very normal life in others. Your grandparents' nominal milestones are not your milestones. You don't need to mourn that. You need to price it in.

A final way to sanity-check your plan is to separate two questions you might be accidentally blending:

- "Is my number growing?" (nominal)
- "Is my *life* getting more funded?" (real)

A nominally growing portfolio can fund a shrinking lifestyle if inflation runs hotter than your plan. A nominally flat portfolio can still represent progress if inflation is low and your savings rate is strong. Your financial story changes depending on which question you ask.

*Actionable takeaway:* Rewrite one big goal in today's dollars and commit to tracking it that way. For example, replace "I want $2 million at 65" with "I want the spending power of $80,000 per year in today's dollars for life" (or whatever fits your life). Then choose one planning method and stick to it: either (a) use a *real* return assumption for your growth bucket, or (b) use nominal returns *and* an explicit inflation rate. The win is not perfect forecasting; it's measuring progress with a yardstick that matches reality.

## 1.4 The Language of Returns

Investment marketing has one favorite word: *average*. Average return. Average annual performance. Average since inception. It sounds calm, statistical, adult. The problem is that your future doesn't live in the average. Your future lives in the sequence of what actually happens—in the order of gains and losses, in the potholes, in the years you'd rather forget. A portfolio isn't graded on the mean of its report card. It's graded on the final GPA.

There are two different "averages" hiding under the same casual phrasing, and confusing them is one of the most expensive category errors in finance. The first is the *arithmetic average*: add the yearly returns and divide by the number of years. The second is the *geometric average*: the single steady annual rate that would have produced the same final result when compounded. The arithmetic average is the average weather. The geometric average is the climate your money actually lived in.

The arithmetic average is simple:

$$\bar{r}_{\text{arith}} = \frac{1}{n}\sum_{t=1}^{n} r_t.$$

If you earned $+10\%$ one year and $-10\%$ the next, the arithmetic average is $0\%$. That feels like "you broke even."

But your account didn't break even. Start with $100:

$$100 \times 1.10 \times 0.90 = 99.$$

You lost $1\%$. The difference is not a rounding error; it's the core mechanic of compounding. Gains and losses multiply, they don't add. Your portfolio speaks multiplication. The arithmetic average speaks addition. That is why so many performance conversations sound fluent while smuggling in nonsense.

The geometric average is the translator:

$$\bar{r}_{\text{geo}} = \left(\prod_{t=1}^{n}(1 + r_t)\right)^{1/n} - 1.$$

In the $+10\%$, $-10\%$ example, the geometric average is negative, because the product is $0.99$. It's the rate your dollars actually experienced.

If this feels pedantic, consider what happens when volatility gets larger. A portfolio that goes up $50\%$ and down $50\%$ does not average to $0\%$ in any useful sense. It's down $25\%$:

$$100 \times 1.50 \times 0.50 = 75.$$

That is why the "math of loss" is so unforgiving. A $-50\%$ drawdown requires a $+100\%$ gain to get back to even:

$$(1 - 0.50) \times (1 + 1.00) = 1.$$

Losses create a deeper hole than they appear to, because the recovery is measured from a smaller base. When people say, "I'll make it back," the hidden question is: *make it back how, and from what base?*

The term for this phenomenon—the gap between the arithmetic story and the actual compounding story—is *volatility drag*. Volatility drag is not a moral judgment about bumpy markets. It's a mechanical description: variability reduces compound growth, even when the arithmetic average return looks fine.

A quick approximation that's surprisingly useful is:

$$\text{Geometric return} \approx \text{Arithmetic return} - \frac{1}{2}\sigma^2,$$

where $\sigma^2$ is the variance of returns (think: the square of volatility, in decimal terms). You don't need to compute it precisely to absorb the lesson: more wiggle in outcomes tends to shave the compounding rate.

The emotional trap is that volatility feels like action, and action feels like opportunity. A stock that swings $5\%$ a day feels like a live wire. A boring diversified portfolio feels like watching paint dry. The market is happy to sell you stimulation. It charges in the currency of long-run outcomes.

Here's a concrete illustration that tends to change how people read performance charts. Imagine two portfolios, both advertised as having the same "average" return over four years: $+10\%$ per year. Portfolio A gets there with calm consistency: $+10\%, +10\%, +10\%, +10\%$. Portfolio B gets there with drama: $+40\%, -20\%, +40\%, -20\%$. Same arithmetic average:

$$\frac{40 - 20 + 40 - 20}{4} = 10.$$

Wall Street can print "10% average" on both brochures. Your bank account will not cooperate.

Compute the ending wealth from $100:

$$\text{A: } 100(1.10)^4 \approx 146.41.$$

$$\text{B: } 100(1.40)(0.80)(1.40)(0.80) = 100 \times 1.2544 = 125.44.$$

That's not a subtle difference. Portfolio B ends with about 14% less money after only four years, despite the same arithmetic mean return. Stretch that gap over decades and the "same average" claim becomes dark comedy.

The point isn't that volatility is always bad. The point is that volatility has a cost, and the cost is paid in the one place investors care about most: terminal wealth. If you're going to take volatility risk, you want to be paid for it with higher expected returns, and you want to be sure you can stay invested through the ugliness that volatility implies. Otherwise, you bought the bumpy ride and skipped the destination.

The diagram below captures the idea visually: two paths, same headline "average," different ending.

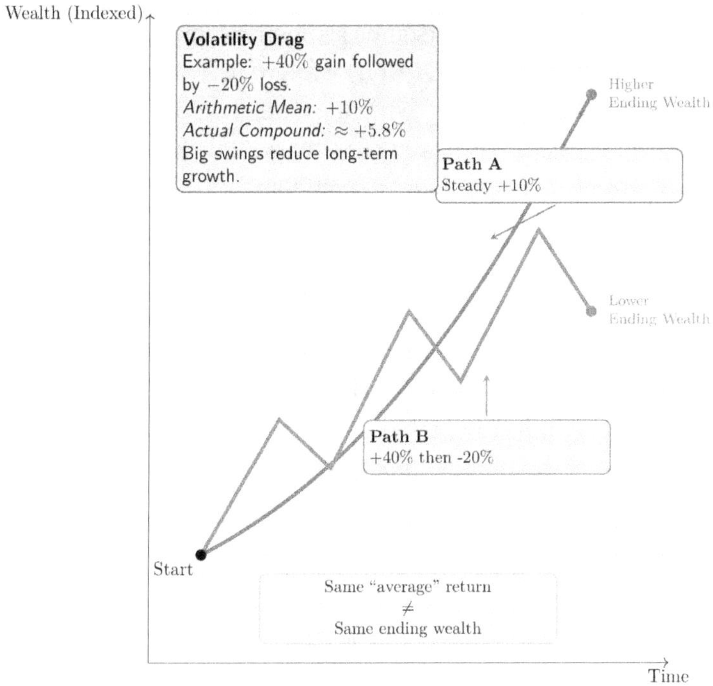


Once you see volatility drag, you start noticing it everywhere. It's why two funds with the same long-run arithmetic average can produce different retirement outcomes. It's why an investor who constantly switches between "hot" assets can end up with a depressing long-term result despite brag-worthy individual trades. It's why boring diversification can beat exciting concentration, even when the concentrated bet has moments of glory.

It also changes how you should interpret the phrase "expected return." Expected return is a statistical average across many possible futures. Your life gets one path. If the return distribution is wide, the average outcome is not the common outcome; it's just the center of mass. That means the *range* of outcomes matters. Volatility doesn't just reduce compounding; it increases the chance you'll experience a path that tests your ability to stick with the plan.

This is where the language of returns meets the psychology of returns. A person who understands geometric compounding but cannot tolerate drawdowns still won't capture it. Your portfolio's compounding rate is not merely what the market delivers. It's what you can hold through. The true return is behavioral: the return after you subtract the cost of your own interventions.

There's another layer of language that quietly misleads people: the difference between CAGR and "average annual return." CAGR—compound annual growth rate—is essentially a geometric measure. It answers: "If my portfolio grew at a constant rate each year, what rate would match the actual start and end values?" It's a useful compression of a messy path into one number, but it can also hide the violence of the ride. A smooth-looking CAGR can come from a stomach-churning sequence.

So how should you talk about returns like a serious adult, not like a brochure?

First, when someone quotes an "average return," ask what kind. Arithmetic? Geometric? Over what period? Before fees or after? Nominal or real? These questions aren't pedantry; they're basic translation. If you don't ask them, you're letting someone choose the language that makes their product look best.

Second, when comparing two investments, don't compare returns without comparing volatility and drawdowns. If one portfolio is smoother, it may deliver a higher geometric return for the same arithmetic headline simply because it avoids deep holes. Avoiding deep holes is not timid. It's mathematically powerful.

Third, remember that taxes and fees are return language too, because what you *keep* is what compounds. A fund's published return may be pre-tax. Your life is not. A strategy that creates frequent taxable gains can have a high stated return and a lower lived return. The IRS doesn't care about your arithmetic average; it cares about realized gains in calendar time.

This is the part where investors often get cynical: "So every number is misleading?" Not quite. The numbers are fine. The language is sloppy. Finance has a translation problem, and you're solving it.

Here's a practical way to reframe performance conversations in your own life: stop asking "What's the average return?" and start asking "What's the likely range of outcomes, and what's the worst path I can tolerate without changing my plan?" The first question invites marketing. The second invites planning.

Also, whenever you're tempted by a strategy promising a higher return, do one brutal check: is the higher return coming from a real edge, or from taking a ride you're unlikely to stay on? If the strategy involves concentrated bets, leverage, or constant trading, the more relevant question is not the expected return. It's the expected behavior. Many investors don't fail because they picked the wrong assets. They fail because they picked assets that turned them into someone they don't like being.

Volatility drag also offers an unexpected kind of relief. It means you don't have to chase heroic returns to win. You can win by building a portfolio that compounds efficiently—a portfolio that avoids huge losses, keeps costs low, stays tax-aware when possible, and matches your risk exposure to your ability to endure it. That's not flashy. It's how long-duration wealth is actually built.

*Actionable takeaway:* The next time you evaluate an investment (a fund, a strategy, or your own portfolio), write down two numbers instead of one:

- The CAGR (geometric return) over the period you care about.
- The maximum drawdown over that same period (the worst peak-to-trough decline).

If you can't find both, you don't yet understand the ride. Then ask yourself one question that cuts through hype: "Could I hold this through a drawdown of that size *without* selling?" If the honest answer is no, the advertised average return is not your return.

## 1.5 Probability and the Nature of Uncertainty

If you've ever asked, "What will the market do next year?" you weren't being naive. You were being human. The brain is a prediction machine. It wants a single number it can grip like a handrail: 8% returns, 2% inflation, retire at 62. One path. One script. One reassuring line on a chart.

The trouble is that markets don't hand out scripts. They hand out distributions—a cloud of plausible outcomes with fat tails, long droughts, occasional fireworks, and plenty of years that feel like nonsense in hindsight. When you demand a single-number forecast, you don't get certainty. You get the *illusion* of certainty, which is often worse because it encourages you to build a plan that can't survive reality.

A serious investor doesn't stop planning. They stop pretending planning is prediction.

Probability is the language that makes this possible. It's not a fancy overlay; it's the only honest way to talk about the future. Instead of asking, "What will happen?" you ask, "What could happen, how often, and what would I do if it does?" That shift sounds philosophical until you realize it changes your portfolio decisions, your savings behavior, and your ability to stay invested when the news gets theatrical.

Start with a simple, almost insulting observation: the "average" outcome is rarely the lived outcome. In a room of 100 adults, the average height might be 5'9". That doesn't mean most people are 5'9". It means some are taller, some are shorter, and the average is a summary statistic that sits in the middle like a polite placeholder. Markets behave the same way. An average return is not a promise. It's a center point that actual years orbit, sometimes politely and sometimes violently.

This is why investors get blindsided by perfectly normal market behavior. They hear "stocks return about 10% on average" and then experience −18% and feel cheated. But −18% is not a betrayal of the average; it's a feature of the distribution. The average is what you get after you blend good years, bad years, and weird years into one smoothie. Your life drinks the years one by one.

The technical word for "how spread out outcomes are" is *variance*. The everyday translation is: how much the ride shakes. Two investments can have the same expected return and very different variance. The expected return tells you where the dart tends to land. Variance tells you how often it lands in your foot.

This matters because the compounding engine doesn't simply care about return. It cares about the path, the magnitude of drawdowns, and your ability to remain invested. A distribution with the same average but wider spread increases the chance of landing in a bad sequence—the kind that tempts you to quit or forces you to sell for life reasons. When people say "I can handle risk," what they often mean is "I can handle risk in a normal year." Markets, with a kind of dark humor, occasionally refuse to give you a normal year.

It also matters because forecast error compounds. If you project wealth 30 years out, a small error in assumed return becomes a big error in ending value, because the error itself gets multiplied year after year. The spreadsheet isn't lying to you; it's doing exactly what exponential math does. It's amplifying whatever assumptions you feed it. The danger is not the spreadsheet. The danger is taking its clean line too seriously.

An honest plan therefore needs an uncertainty statement. Not a hedge like "markets are uncertain," but a usable range: "If returns are weak for a long time, I can still pay my bills and keep contributing; if returns are strong, I won't inflate my lifestyle so quickly that I ruin the advantage." This is not pessimism. It's engineering. Bridges are built with safety margins not because engineers expect the bridge to fail, but because they respect wind.

A powerful mental move is to stop treating your plan as a single forecast and start treating it as a set of scenarios. Scenarios don't require you to know the future. They require you to acknowledge the menu of futures the market tends to offer.

A practical scenario set might look like this:

- *Base case*: returns and inflation near your long-run assumptions.
- *Bad-luck decade*: low or negative real returns early on, plus life expenses that don't care.
- *Good-luck decade*: unusually strong returns that tempt you to relax too soon.

This isn't a "list of three" for aesthetic reasons. It's the minimum set that forces you to confront the two ways plans break: discouragement and complacency.

The discouragement case is obvious: a long flat market makes you feel like the engine is broken. The complacency case is sneakier: strong early returns make you feel like you were right, like you have a gift, like you can ease off saving because the portfolio "has it." Then the market mean-reverts and your plan reveals it was depending on a run of luck.

If you want an emotionally accurate definition of risk, try this: risk is the chance you will meet yourself under stress and not like what you do. The math version is volatility and drawdowns. The human version is panic selling at the bottom, leverage at the top, and the sudden desire to "wait until things calm down" right before the rebound.

Probability thinking doesn't eliminate those impulses. It anticipates them. It builds guardrails that don't depend on willpower.

One way to make distributions intuitive is to picture a jar of marbles, each marble representing a yearly return. Some marbles are green (good years), some are red (bad years), most are boring beige (unremarkable years). Your lifetime investing experience is not one marble. It's a handful pulled in a specific order. The average return is like saying, "On average, marbles are slightly green." That tells you almost nothing about the next pull, and less than nothing about whether you get three red marbles in a row early in retirement.

Order matters because compounding multiplies. The same set of returns can produce different outcomes depending on sequence, especially when you're withdrawing. Even before withdrawals, sequence matters psychologically: a bad first few years can convince you the plan is foolish, even if the long-run average would have rewarded you.

This is why long-horizon investors should care about *range* as much as they care about *center*. A range forces humility. It also creates better decisions. If your plan only works in the base case, it isn't a plan. It's a wish with formatting.

So what should you do with uncertainty besides admire it?

First, attach probabilities to things that have deadlines. If you need a down payment in three years, your question is not "What return will stocks deliver?" It's "What is the probability that stocks will be down at the exact wrong time, and can I tolerate that outcome?" Most people, when they answer honestly, end up moving deadline money out of the growth engine. That is not market timing. That is matching the tool to the job.

Second, for long-duration money, stop optimizing for a single number and start optimizing for *staying invested*. The best "expected return" in theory is worthless if it produces a distribution of outcomes that causes you to abandon the strategy. The investable strategy is the one you can actually hold. Probability thinking turns this into a design constraint: choose an allocation whose bad-case path you can survive emotionally and financially.

Third, run sensitivity analysis. That's a sterile term for a healthy habit: ask "What if I'm wrong?" If you assume a 5% real return and reality delivers 3%, what changes? Retire later? Save more? Spend less? If none of those are acceptable, then the plan is not robust; it's fragile. Better to discover fragility in a spreadsheet than in a market crash.

A good sensitivity analysis doesn't require complicated software. You can do it with three return assumptions and three inflation assumptions. The goal is not to pin down the truth. The goal is to see which levers dominate. For most households, a higher savings rate and a longer timeline matter more than pretending you can forecast returns to the second decimal place.

Probability thinking also cures a specific and common kind of overconfidence: the belief that your personal experience is representative. If you started investing in a bull market, you may feel markets are generous and quick to recover. If you started in a flat decade, you may feel markets are stingy. Both experiences are real. Neither is the full distribution. The market is larger than your sample size.

The same bias shows up with inflation. If you lived through a period of low inflation, you may plan as if inflation is a rounding error. If you lived through a spike, you may plan as if it's permanent. Probability thinking doesn't force you into one assumption; it forces you into ranges and contingencies.

There's also a social-media version of this bias: "If you had just bought X in year Y, you'd be rich." That statement is usually true and almost always useless. It picks one realized path and ignores the probability distribution of alternative paths, including the chance you would not have held through drawdowns or would have sold too early. In the real world, the hardest part is not discovering the winning asset. It's holding a reasonable strategy long enough for compounding to do its slow work.

You don't need to become a statistician to become probabilistic. You need one habit: when you see a point estimate, mentally ask for a confidence interval. When someone says, "This investment should return 8%," translate it to: "It might return -30% in a bad year, it might return +30% in a great year, and over decades it might average somewhere in the neighborhood of 8%—if I can hold it." That translation makes you harder to manipulate and less likely to manipulate yourself.

Here's a concrete way to apply this to your own planning without building a model worthy of a hedge fund. Pick a target outcome—say, a portfolio value in today's dollars, or a level of spending power. Then calculate three paths:

- *Conservative*: lower real return than you hope.
- *Middle*: your best guess.
- *Optimistic*: higher real return than you hope.

Now look at the differences and ask: which levers could close the gap if reality is conservative? If the answer is "save an extra $300/month" or "work one more year," you have robustness. If the answer is "I guess I'd need a miracle," you have fragility dressed up as a plan.

The final step is emotional, not mathematical: accept that uncertainty is not a personal insult. When markets are uncertain, they are not singling you out. They are being markets. The investor who internalizes uncertainty as failure tends to react, trade, and tinker. The investor who treats uncertainty as physics tends to build slack, diversify, rebalance, and keep funding the engine.

*Actionable takeaway:* Rewrite your plan using ranges instead of single numbers. For your long-duration portfolio, choose a conservative, middle, and optimistic *real* return assumption (after inflation), and compute what each implies for your goal. Then write one sentence for each: "If the conservative case happens, I will do ___." Your answer should be a lever you control (save more, spend less, extend the timeline, reduce required spending power), not a lever you don't ("find the next winning stock"). If you can't write a credible sentence for the conservative case, you don't have a plan yet—you have a forecast.

## 1.6 Fueling the Machine: Cash Flow and Slack

The most dangerous sentence in personal finance is, "I'll start investing more when things calm down." For most households, things never calm down. They just change costumes. One month it's a surprise dental bill; the next it's a car that develops a personality; then it's a rent increase that arrives like a landlord's annual tradition. The compounding engine doesn't stall because markets are mean. It stalls because life is loud.

Cash flow is the part of wealth-building that feels least glamorous and ends up being the most decisive. It's not about spreadsheets; it's about oxygen. When your monthly cash flow is tight, every market dip becomes a potential emergency, because you have no slack—no room to absorb shocks without interrupting your plan. When your cash flow has slack, volatility turns from threat into background noise.

Most investors obsess over expected returns because returns are exciting. Savings rates are not. Savings rates sound like scolding. But the savings rate is the one lever that doesn't pretend to be probabilistic. You control it. You can't control what stocks do next year, but you can control whether you keep fueling the machine while they do it.

A useful definition: your savings rate is the percentage of your income you *keep* for your future self. Not what you earn. Not what you invest in. What you keep. That sounds simple until you realize it's also a statement about identity. Every dollar you save is a vote for future freedom; every dollar you spend is a vote for present comfort. Neither vote is immoral. The trap is voting without noticing you're voting.

Financial slack is what makes those votes stable. Slack is the margin between your life and your limits: extra cash, extra time, extra flexibility. It's the difference between a plan that works in the spreadsheet and a plan that works when your boss gets weird, your kid gets sick, or your roof leaks in the same week your index fund decides to have a tantrum.

Slack is not the same thing as being "rich." You can have a high net worth and no slack if all your wealth is illiquid and your spending is maxed out. You can have a modest net worth and meaningful slack if your expenses are low relative to your income and you keep cash reserves. Slack is not a status symbol. It's a system property.

Here's a picture you may recognize. An investor builds a growth portfolio and feels proud—until a real-world shock hits. They lose a job, or hours get cut, or a family emergency appears. Suddenly the growth portfolio is no longer "long-duration money." It becomes the only available pool of dollars. They sell. The market happens to be down. They lock in losses. Later they return to investing, but the base is smaller and their confidence is cracked. They blame the market.

That wasn't a market failure. It was a fuel system failure.

Long-duration wealth is built by doing something that looks almost boring: continuing to add money when you don't feel like it, and continuing to hold through periods that make you question your intelligence. The only way most people can do that is if their day-to-day finances are not a knife-edge.

Slack is how you buy the ability to be patient.

One of the best ways to understand slack is to stop thinking of it as "cash that could have been invested" and start thinking of it as *insurance against being forced to make the worst possible decision at the worst possible time*. The market has its own schedule. Your bills have their own schedule. Slack is what prevents those schedules from colliding.

A household with no slack becomes a forced seller. A forced seller doesn't get to choose when to realize losses. They get chosen. That is a terrible negotiating position, and the market is a ruthless negotiator.

So what does slack look like in practice?

First, it looks like an emergency fund sized to your life, not to a rule of thumb you saw on the internet. "Three to six months of expenses" is a starting point, not a commandment. The right number depends on your income stability, the number of dependents, the health of your car, the age of your house, and how quickly you could replace your job. If you are dual-income in stable fields, your required slack may be lower. If your income is variable or you carry high fixed costs, you need more shock absorption.

Second, slack looks like low fixed costs relative to income. Fixed costs—housing payments, car payments, subscriptions you forgot about, debt service—are the part of your budget that refuses to cooperate when life changes. High fixed costs turn a small income disruption into a crisis. Lower fixed costs turn the same disruption into an inconvenience. This is why people who keep their lifestyle "light" often invest more consistently than people who make more money but spend like they're racing themselves.

Third, slack looks like the ability to handle a $400 surprise without touching investments. That number is not sacred; it's just a clean test of fragility. If a few hundred dollars can derail you, the compounding engine is operating without a safety valve. You are not failing morally. You are operating a machine without basic containment.

The deepest insight about savings is that it is not primarily an act of deprivation. It is an act of *control.* Investors often think the market is the main driver of their outcomes. Over a decade, maybe. Over a lifetime, your cash flow system decides whether you can stay invested, keep contributing, and avoid forced selling. Those behaviors create the conditions under which the market's long-run rewards can actually reach you.

You can see this in two investors with the same portfolio allocation. Investor A saves 5% of income and has no cash buffer. Investor B saves 20% and keeps an emergency fund. In a bull market, both feel fine. In a bear market, Investor A is stressed and tempted to sell; a job disruption becomes catastrophic; contributions stop. Investor B keeps contributing and may even rebalance into cheaper assets. The difference in outcome is not intelligence. It's fuel and slack.

Slack also changes how you experience volatility. Without slack, volatility feels like a threat to your rent. With slack, volatility feels like what it is: a normal feature of markets. It's hard to overstate how much this changes behavior. Many investors think they need more courage. Often they need more cash reserves.

There's another way slack feeds compounding: it gives you *optionality*. Optionality is the ability to make a choice rather than accept a default. With slack, you can take a better job rather than the first job. You can move. You can negotiate. You can take a course, start a business, or step away from a toxic situation. Those choices can increase income, reduce stress, and indirectly increase your savings rate. Slack is not just defensive; it's offensive. It's career leverage.

This is why the old "budgeting" conversation is often mis-framed. People treat budgeting as a punishment for spending. The more useful framing is that budgeting is how you design a cash flow system that protects your growth bucket from being raided. You are not tracking every latte because you're a monk. You are tracking cash flow because you want your compounding engine to run uninterrupted for decades.

A practical system has two parts: automation and friction.

Automation means you remove the need to decide every month. Contributions happen first. Bills get paid. Spending adapts to what remains. This is the opposite of the common approach, where spending happens first and investing gets whatever scraps are left over. Scraps do not build long-duration wealth.

Friction means you make it slightly harder to sabotage yourself. If your spending account and your investment account are one click apart, you have built a temptation pipeline. People who consistently build wealth often add small, boring barriers: separate accounts, automatic transfers on payday, and a rule that any change to contributions requires a 72-hour waiting period. That waiting period is not about discipline; it's about respecting that you are a different person when you're stressed.

Savings rates feel abstract until you connect them to what they purchase: time. A higher savings rate doesn't just buy a bigger account balance. It buys the ability to take market risk without panic, because you are constantly adding fuel. It buys the ability to recover from bad sequences. It buys the ability to retire earlier or work less or be pickier. In a very literal sense, your savings rate is how much of your future you are buying every month.

There's a curious irony here: people who feel behind often try to fix it by taking more investment risk. More volatile assets, more concentrated bets, more cleverness. Sometimes that works temporarily. Often it fails spectacularly. The more reliable way to fix being behind is to increase the fuel flow. That's not because returns don't matter. It's because the combination of higher savings plus time creates a stronger engine than a desperate search for a miracle return.

This is also where tax-advantaged accounts matter. They are not just account types; they are higher-capacity fuel lines. If you can put more dollars into vehicles that shelter growth from taxes, you increase the portion of return that remains in the engine to compound. Contribution limits are essentially the system telling you how much fuel you're allowed to inject into the tax-sheltered chamber each year. If you are in a position to use that capacity, it's one of the cleanest ways to improve long-run outcomes without changing the investment strategy at all.

But beware the emotional trap: "I'll max out the account" can become another vague ambition that never lands. The better approach is to treat contributions as an engineering ramp. Increase them gradually. Tie increases to raises. If you get a 5% raise, capture 2% for your future self before your lifestyle learns it exists. Lifestyle inflation is just inflation of a different kind: it erodes future freedom by making present spending feel non-negotiable.

Slack-building also pairs beautifully with probabilistic planning. When you accept that bad sequences happen, you stop treating cash reserves as wasted return. You treat them as an explicit part of the plan. The reserve is what allows you to keep investing during bad markets and bad life periods, which is exactly when the future expected return is most valuable. It is not a coincidence that investors with steady contributions through bear markets often end up with surprisingly strong long-term results. They bought more shares when shares were cheaper, not because they were brave, but because they were funded.

If you want one sentence that captures the relationship: slack turns uncertainty from a portfolio problem into a calendar problem you can handle.

That sentence also explains why "high income" is not enough. High income without slack often produces high spending, which produces fragility, which produces reactive decisions, which produces poor compounding. Meanwhile, moderate income with strong slack can produce consistent investing and an investor who stays rational. The market rewards consistency more reliably than it rewards cleverness.

A useful exercise is to compute your *financial runway*: the number of months you could cover your cannot-fail expenses if your income dropped to zero. This is not a fear exercise. It's a power exercise. Runway tells you how much time you have to make good decisions. Without runway, you make fast decisions. Fast decisions are often expensive.

Once you have runway, you can treat your savings rate like a dial. You can adjust it up or down based on life stage, but you don't turn it off. Turning it off is how people lose decades. Even a small continuing contribution keeps the habit alive and keeps you connected to the compounding engine.

If life is tight right now, the advice is not "save 30%." That's a fantasy that produces guilt. The advice is "build the first inch of slack." The first inch might be $500 in a separate account. Then $1,000. Then one month of essentials. The early stage is not about maximizing return. It's about reducing fragility. Fragility is what keeps people from ever reaching the stage where compounding feels powerful.

*Actionable takeaway:* Measure your slack and set a concrete fuel rule. This week, calculate (1) your monthly "cannot-fail" expenses and (2) the cash you could access without selling long-duration investments. Divide (2) by (1) to get your runway in months. Then pick a simple operating target: "I will build runway to ___ months, and I will automate a savings rate of ___% (or $___/month) that increases by ___% of every raise." The number doesn't need to be perfect. It needs to exist. A compounding engine with a steady fuel line and a safety buffer is not just more profitable; it's harder to break when life gets loud.

Chapter 2

# The Return Drivers: Asset Class Mechanics

*If you view a stock ticker as just a wiggling line on a screen, you are gambling, not investing. To build wealth that lasts decades, we need to pop the hood and understand exactly where returns come from. This chapter strips away the jargon to reveal the mechanical engines inside equities, bonds, and real assets-showing you precisely how they generate cash for your future self, and why they sometimes stall.*

## 2.1 The Palette: Asset Classes Overview

Picture two investors watching the same headline scroll by: "Inflation surprises to the upside." One shrugs—his portfolio barely flinches. The other feels a small jolt of dread, because he already knows what his account statement will look like next month. The difference isn't bravery or intelligence. It's that they're holding different *economic jobs* inside their portfolios.

Most people learn investing by *labels*: "tech stocks," "emerging markets," "real estate," "bonds." Labels are convenient, but they're also a little like judging a restaurant by its signage. The thing that matters is what's happening in the kitchen: what cash flows are being produced, how those cash flows change with the economy, and what you're secretly betting on when you hold the asset. If you organize your portfolio by economic function, the chaos of "asset classes" turns into a manageable palette-like a painter choosing colors based on what mood and lighting they need.

A useful mental model is to separate investments into three functional buckets: *engines* (capital appreciation), *anchors* (capital preservation), and *shields* (inflation protection). These aren't moral categories-no asset is "good" or "bad." They're roles in an ecosystem. Engines help you grow wealth. Anchors help you *not get knocked out of the game*. Shields help you keep your purchasing power when money itself is quietly losing value.

The sneaky point: the same ticker can behave like a different color depending on *when* you use it and *what price* you pay. A long-term Treasury bought at a 1.5% yield is a very different animal than the same bond bought at 5%. A "safe" dividend stock bought at an absurd valuation can act less like an anchor and more like a fragile engine that stalls at the first pothole. Function is not a permanent identity; it's a relationship between the asset and the economic environment.

### Engines: claims on growing cash flows

Engines are assets that can plausibly grow your purchasing power faster than inflation over long stretches, not because of magic, but because they sit on top of something that can scale: a business, a set of properties with rising rents, an economy that becomes more productive. The defining trait is not day-to-day volatility. It's that, over time, the underlying cash flow base can expand.

Public equities-stocks-are the cleanest example. You own a slice of a business that sells things, pays people, competes, innovates, and (ideally) earns profits. That profit is the raw material for compounding. Some gets paid out to owners; some gets reinvested to make the future business bigger. When everything works, your wealth grows because the machine underneath your shares grows.

It's tempting to say "stocks return about X%," but engines don't come with a guaranteed speed. An engine can idle for years. It can backfire. It can be overpriced such that future returns are pulled forward into the present, leaving the next decade to digest the bill. The reason engines are powerful isn't that they're always comfortable—it's that they're one of the few ways an individual investor can harness the long-run tendency of human systems to improve: better tools, better logistics, better medicine, better software, better management. Progress is messy, but over decades it tends to show up in the cash flows of widely owned businesses.

Real estate-especially in diversified, professionally managed form-often sits in the engine bucket too. A rental property is not a "house"; it's a contract with tenants mediated by local wages, local supply, and local regulation. Rents can grow. Replacement costs can rise. Leverage can amplify results (for better and worse). Over time, real estate can behave like an engine *with a cash yield*, which is psychologically comforting: you see money arriving while you wait for the long game to play out.

Private businesses and private equity are engines as well, but with an added twist: they require you to accept less liquidity and less transparency in exchange for the chance to buy and improve cash-flowing assets away from the public spotlight. The economic function is still the same-ownership of an operating system that can throw off and grow cash.

A critical emotional feature of engines: they test your identity. When stocks fall 30%, the market isn't just repricing future cash flows. It's also whispering that you're naive for believing in the long run. Engines are where most compounding comes from—and where most self-sabotage occurs—because the path is jagged, and humans are wired to confuse volatility with danger.

## Anchors: contracts that stabilize the boat

Anchors exist for a simple reason: you will be asked to make decisions at the worst possible times. If every asset you own can drop together, your portfolio becomes a psychological trap. The anchor bucket is built from assets that tend to be more stable in nominal terms and, crucially, tend to hold up (or even improve) in some of the environments where engines are suffering.

High-quality bonds-especially government bonds-are the classic anchors. A bond is an IOU: you lend money, and you are promised a schedule of interest payments and a return of principal. That contractual structure is the point. You're not hoping management makes great decisions; you're renting out your money for a stated rate.

Cash and cash equivalents (like Treasury bills and money market funds that hold short-term government or high-quality instruments) are the simplest anchor of all. They don't promise high returns. They promise *optionality*. Cash gives you the ability to act when others can't: to pay bills, to rebalance, to buy assets on sale, to sleep. In investing, the ability to stay solvent and sane is an edge-and cash is often the cheapest tool for buying that edge.

Anchors have an unglamorous job description, which is why investors constantly mistreat them. People demand that anchors "perform." They complain when safe assets lag in bull markets, as if a fire extinguisher has failed because it didn't produce income. The real test of an anchor is not whether it wins when everything is going up. The test is whether it prevents a bad decade from turning into a bad life decision—panic-selling a long-term portfolio to stop the emotional bleeding.

There's also a subtle but important point: anchors are not "no-risk." They are simply different risks. Cash has inflation risk. Longer-maturity bonds have interest-rate risk: when new bonds offer higher yields, older bonds offering lower yields must drop in price to compete. Corporate bonds add credit risk: the company might struggle or default. The anchor bucket is about stabilizing the portfolio's behavior, not eliminating uncertainty from the universe.

If you want a practical way to identify an anchor, ask a harsh question: "If the economy stumbles and investors become fearful, does this asset become more attractive or less attractive?" High-quality government bonds often become more attractive in fear regimes because investors rush toward contractual safety. Highly leveraged companies, speculative credit, and economically sensitive assets often become less attractive at the same time. That difference is diversification in its most useful form: not many colors, but different reactions to the same light.

### Shields: assets that fight purchasing-power erosion

Shields exist because inflation is a quiet form of confiscation. It rarely arrives with drama. It arrives with a slightly higher grocery bill, a rent renewal you don't love, a "reasonable" price increase that somehow happens every year. Over decades, inflation can turn "safe" nominal wealth into a shrinking real lifestyle.

The key idea is that some assets have cash flows or principal values that can *reprice* when the currency reprices. That repricing can be explicit, contractual, or implicit. Explicit repricing is the cleanest: inflation-linked bonds such as U.S. Treasury Inflation-Protected Securities (TIPS) adjust principal with the consumer price index. That makes them feel almost like a financial product designed by someone who got tired of pretending inflation is a minor detail. Their job isn't to make you rich; it's to make sure your "safe" money stays safe *in real terms*.

Commodities, in broad diversified form, can behave like a shield in unexpected inflation shocks because they are closer to the raw inputs of the economy. When the price of energy, metals, or agricultural goods rises sharply, it's often *those very prices* that are driving inflation headlines. That said, commodities are a high-maintenance shield: they can be volatile, they can go through long stretches of disappointing real returns, and they require a structure (like futures-based funds) that introduces its own mechanics. A shield isn't automatically comfortable.

Gold is the most psychologically charged shield. It produces no cash flow. It just *sits there*. And yet, across history, it has repeatedly served as a kind of financial "spare tire" for regimes where people lose trust in paper promises. For a modern investor, gold's value is less about everyday inflation and more about the tails: currency debasement fears, geopolitical shocks, and moments when confidence in institutions wobbles. It's a hedge against certain narratives—and narratives can move markets.

Real estate can also act as a partial shield because rents and replacement costs can adjust with inflation over time. But here the "function not label" rule matters. A property with long leases and capped rent escalators is different from apartments with annual repricing. A REIT with inflation-sensitive cash flows is different from a REIT whose tenants are struggling retailers. Real estate is not a single asset; it's a bundle of local economies, lease structures, financing terms, and political constraints.

Even equities have some shield-like qualities over long horizons because businesses can raise prices, cut costs, innovate, and adapt. But "stocks hedge inflation" is only true in the way "exercise improves health" is true: over time and with consistency, not necessarily in the moment you most want relief. In inflationary shocks, stocks can suffer if higher rates compress valuations or if consumers pull back. Engines can contain shield components, but they are still engines: they can stall.

### The environments that matter (and why investors keep being surprised)

Asset allocation becomes far less mysterious when you stop thinking in "markets" and start thinking in "environments." The environments that do the most damage to investors aren't the ones that are merely unpleasant; they're the ones that are *unexpected*. The market is a machine for pricing what most people already believe. Your portfolio gets tested by what most people didn't price in.

One useful way to think about environments is to imagine two axes: *growth* (strong vs. weak) and *inflation* (high vs. low). You don't need a PhD to see why this matters. Strong growth usually helps engines because cash flows expand. Weak growth stresses engines because profits and optimism shrink. High inflation stresses nominal anchors because the currency is being diluted. Low inflation (or disinflation) can help high-quality bonds, because yields can fall and existing bonds become more valuable.

In strong growth with contained inflation, engines tend to look heroic. Companies hire, invest, and report rising earnings. Credit is plentiful. Risk feels like an abstract concept. The trap is that investors then treat engines as if they are anchors—levering up, concentrating, and assuming the good environment is a permanent feature of reality.

In weak growth with low inflation (or deflation scares), high-quality bonds can shine as anchors. When investors fear recession, they often accept lower yields for safety, pushing bond prices up. This is the environment where many people first learn that "safe" assets can have surprisingly good returns—at exactly the moment that "risky" assets feel personally insulting. The trap here is symmetric: investors decide the engine is "broken" and try to live off anchors forever, locking in low expected returns right after an attractive entry point appears for growth assets.

In strong growth with high inflation, life gets weird. Companies can raise prices, but costs rise too; wage pressure bites; central banks may tighten; and valuation multiples can compress because investors demand higher returns. Some real assets can do well, and businesses with true pricing power can keep up. But it's not a clean party for stocks, and it's often a miserable time for long-duration bonds. The trap is thinking that because "the economy is growing," everything should go up together.

In weak growth with high inflation—stagflation—your portfolio gets asked an uncomfortable question: "What do you own that doesn't need a thriving economy to protect your purchasing power?" This is where shields earn their name. It's also where the investor's urge to chase recent winners becomes most destructive, because the regime itself is confusing and emotionally exhausting.

Notice what this framing does: it changes diversification from "own lots of stuff" to "own things that *win for different reasons.*" Diversification is not a scavenger hunt for more tickers. It's a deliberate attempt to avoid having your financial future depend on a single macro storyline.

### Why "industry diversification" feels good and fails quietly

A portfolio split across healthcare, technology, consumer staples, and financials can look diversified and still behave like a single bet: a bet on economic growth and stable financing conditions. In a true risk-off event, correlations often rise—meaning assets that usually wiggle independently suddenly move together. This is the market's version of gravity: when fear spikes, investors sell what they can, not what they should.

That doesn't mean industry diversification is useless; it reduces company-specific risk. But it often doesn't address the deeper portfolio problem: what happens when the discount rate changes, when inflation surprises, or when credit tightens. Those are environment-level shocks, and they cut across sectors.

A memorable way to test your diversification is to pick a headline you *don't* want to see something like "rates rise sharply," "credit spreads blow out," "inflation resurges," "recession odds jump." Then ask: "Which parts of my portfolio are designed to not care?"

If the honest answer is "none," what you have is a single story told with different characters.

## Putting the palette to work: choosing roles before choosing products

When investors start with products ("Should I buy this ETF?"), they often end up with an accidental portfolio: a pile of exposures that overlap, leave gaps, and react the same way in stress. When you start with roles, product selection becomes almost mechanical.

First, decide how much engine you need. This isn't about how brave you feel today; it's about whether your goals require growth. If you're saving over decades, engines are not optional. If you need the money next year, engines may be inappropriate, not because they're "bad," but because time is the shock absorber that makes engines work.

Second, decide what anchors you need to prevent self-sabotage. The most expensive investing mistake is not "earning 1% less." It's bailing out of engines during a drawdown and never getting back in. Anchors can be thought of as the price you pay to avoid becoming your own worst enemy. The right amount of anchoring is the amount that keeps you from making irreversible decisions when headlines get loud.

Third, decide what inflation protection you need based on your liabilities. A retiree with spending needs tied to groceries, rent, healthcare, and insurance has an inflation-linked problem whether they acknowledge it or not. A young worker with a rising income might have a natural inflation hedge because wages often rise with inflation over time. The portfolio's shields should reflect the life you're trying to fund, not a generic "inflation is bad" slogan.

Once roles are clear, you can be picky about implementation. You can choose broad, low-cost index funds for engines. You can choose high-quality bonds and short-term cash-like instruments for anchors. You can choose explicit inflation-linked instruments (like TIPS) and carefully sized real-asset exposures for shields. The details vary, but the logic stays steady: every holding earns its place by doing a job.

### A concrete way to audit your own portfolio tomorrow morning

Open your portfolio and, holding nothing sacred, assign each position a role: engine, anchor, or shield. If a position can't be assigned a role in one sentence—if it's just there because it's familiar, exciting, or "someone smart likes it"—flag it.

Then ask two questions that cut through almost every rationalization:

- "In a recession scare, what do I expect this to do—and why?"
- "In an inflation surprise, what do I expect this to do—and why?"

If you can't answer without resorting to hope ("I think it'll be fine"), you've found either a knowledge gap or a portfolio gap. Both are fixable, but only if you're honest about them.

### Actionable takeaway

Write a one-page "portfolio job description" before you change a single holding: the percentage you want in *engines* (growth), *anchors* (stability), and *shields* (inflation protection), plus one sentence describing what each bucket is meant to protect you from. Then audit every current holding against those jobs and sell only what can't explain its purpose in plain English.

## 2.2 Equities: The Compounding Engine

The stock market trains people to watch the wrong scoreboard. It flashes prices in neon, updates them by the second, and turns a slow-moving business into something that feels like a video game. If you're not careful, you start treating a share of stock like a scratch-off ticket: you buy it, you wait for the "hit," and you judge the decision by what the quote does next.

A stock is not a lottery ticket. It's a property right.

When you own equity, you own a fractional claim on a living, breathing cash-flow machine-one that hires people, buys raw materials, ships products, argues in meetings, makes mistakes, learns, adapts, and (if it survives) throws off profits year after year. Price is what other people are willing to pay for that claim *today*. Value is the business's ability to deliver cash to its owners over time. Those two things sometimes travel together. Sometimes they take different flights for a while.

The most useful mental shift is to stop thinking "Will this stock go up?" and start thinking "What does this business *do* with the money it earns?" The long-run compounding story of equities isn't primarily about clever trading or perfectly timed entries. It's about capital allocation—how profits get split between money paid out to owners and money kept inside the business to make the business bigger.

## A share is a claim on leftovers---and that's the feature

Debt holders get paid first. Employees get paid before owners. Suppliers get paid. Landlords get paid. The government gets paid. Equity gets what's left—sometimes a lot, sometimes nothing. That "residual" nature is exactly why stocks can compound so powerfully. If the business becomes wildly more productive, equity participates in the upside after the fixed claims are satisfied.

This also explains why equities can feel emotionally rude. A bond feels like a promise. A stock feels like a shrug. When times get tough, a company can cut the dividend, issue more shares, dilute you, or reinvest in survival. Equity is the flexible layer in the capital stack.

Flexibility is risk. Flexibility is also where the long-run reward comes from.

If you want to understand stocks the way professionals do, you don't start with chart patterns. You start with the income statement and the cash flow statement—because that's where you see whether the business is actually producing surplus cash, and what it chooses to do with it. You don't need to become an accountant; you just need the one idea that makes everything click: profit is a fork in the road.

### Net income: the business's ``surplus,'' with strings attached

When a company reports *net income*, it's telling you (in simplified terms) "After paying expenses and taxes, this is what we earned." Net income is not the same thing as cash in the bank-accounting has timing quirks-but it's still the cultural heartbeat of corporate performance. CEOs get judged on it. Boards talk about it. Analysts model it. Markets react to it.

And then comes the part most investors never visualize: net income doesn't sit politely in a pile waiting to be admired. It has to be allocated.

Companies can send profits out to shareholders as dividends. They can also buy back shares, which is another way of returning cash to owners-less like a paycheck and more like the company quietly reducing the number of slices the pie is cut into. Or they can retain earnings, keeping the money inside the business to fund growth, reduce debt, build factories, hire engineers, develop products, acquire competitors, or simply create a cushion for hard times.

That retained portion is the source of *internal compounding*. It's compounding that happens inside the walls of the company, largely invisible day-to-day, like a city upgrading its plumbing while tourists argue about the skyline.

Internal compounding is why long-term equity returns can be so stubbornly resilient even when the investor experience is full of drama. The market may be chaotic, but businesses-good ones-wake up every day and try to earn money again. If they can reinvest a portion of those earnings at decent rates of return, your ownership claim can grow even if you never add another dollar.

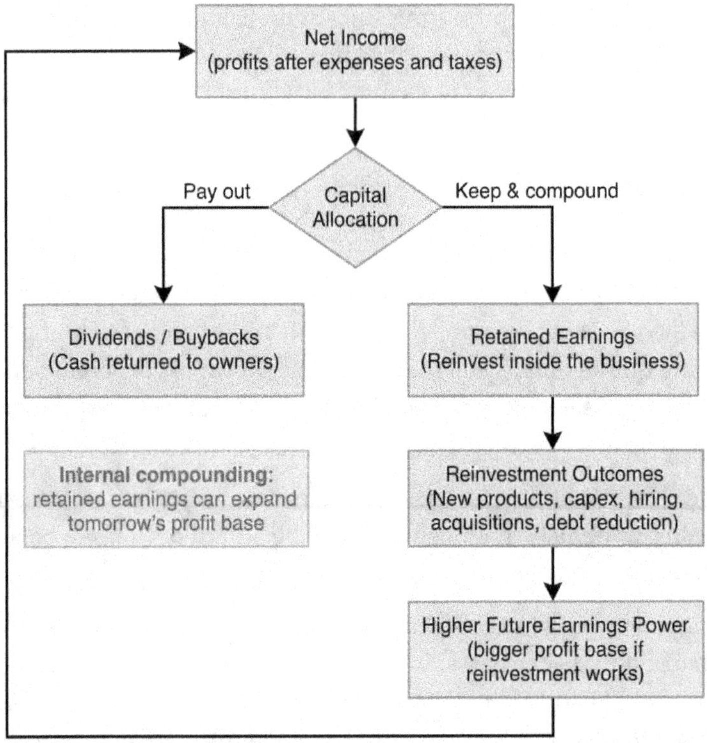


### Dividends are visible. Retained earnings are the iceberg

Dividends feel satisfying because they're concrete. You see the cash arrive. Your brokerage statement congratulates you. You can spend it, reinvest it, or just enjoy the psychological relief of "getting something back."

Retained earnings are less charming. They require trust—and patience—and an eye for whether management is reinvesting intelligently or just building an empire to feel important. But retained earnings are often where the real wealth is made, because they can grow the company's future earning power without you doing anything.

A simple thought experiment makes this visceral. Imagine a small business that earns $1 million a year after expenses. It can either pay you that $1 million every year forever, or it can keep $500,000 of it and reinvest in projects that earn, say, 15% on invested capital. If those projects are real—if they aren't accounting tricks or vanity acquisitions—then next year's earnings aren't $1 million anymore. They're $1 million plus the incremental profit from reinvestment. Repeat this long enough and the profit base can become unrecognizable compared to where it started.

That's compounding in its most practical form: not a formula on a whiteboard, but a business that keeps finding ways to turn today's surplus into tomorrow's larger surplus.

This is also why the quality of growth matters more than the *rate* of growth. A company can grow revenue rapidly by discounting products, buying customers with marketing spend, or acquiring businesses at silly prices. That kind of growth is a sugar high. Compounding growth is funded by real economic surplus and reinvested at attractive returns. It's quieter. It's sturdier. It often looks boring right up until it doesn't.

### The three shareholder payout channels (and why one gets misunderstood)

Shareholders can get paid in three main ways, and two of them are frequently misread:

*Dividends* are direct cash payments. They're easy to track and psychologically reassuring. Mature businesses-utilities, consumer staples, some financials-often use dividends as a discipline mechanism: "If we can't find good projects, we'll hand the cash back."

*Buybacks* are indirect cash payments. When a company buys its own shares, it reduces the share count. If the business's total earnings stay the same, each remaining share represents a larger claim on those earnings. Buybacks are controversial because they can be abused (especially when done at inflated prices or to offset dilution from stock-based compensation), but the basic mechanism is simple: fewer slices, same pie.

*Reinvestment* is payment delayed-sometimes wisely, sometimes foolishly. If reinvestment works, you get paid later through higher earnings, higher dividends, more buybacks, and often a higher stock price. If reinvestment fails, you get neither today's payout nor tomorrow's growth. This is why capital allocation is one of the most important skills in corporate leadership-and one of the most underappreciated risks in equity investing.

Here's the psychological trap: investors often treat dividends as "real return" and everything else as "speculation." In reality, all shareholder returns are ultimately funded by the same source: the business's ability to generate cash over time. Dividends are just the part that shows up with a receipt.

## Why stock prices move like rumors while businesses move like glaciers

If stocks are ownership in businesses, why do prices look so unhinged?

Because the stock market is not a courtroom discovering truth. It's an auction clearing price. And auctions respond to marginal changes in belief. If enough participants decide the future is a little brighter or a little darker, the price moves immediately-even though the factory floor looks exactly the same.

Markets are forward-looking, but "forward-looking" doesn't mean "accurate." It means "constantly repricing the story." Stories get revised for reasons that have nothing to do with long-run compounding: interest rates change, sentiment shifts, geopolitical risk spikes, a competitor announces a product, an analyst publishes a persuasive narrative. The market's job is to incorporate all that into a price *right now*. Your job is to decide whether the price is a sensible deal for an ownership claim that may last decades.

This is why treating equities as engines is a behavioral strategy as much as a mathematical one. The math says compounding needs time. The psychology says time is hard because you have to live through noise. Seeing a stock as a business helps because it moves your attention from "How do I feel about the quote?" to "How is the engine doing?"

A useful habit is to periodically translate your holdings into plain English ownership statements:

> "I own a tiny piece of a global payments network," or "I own a claim on thousands of apartments," or "I own a slice of companies that sell everyday products and raise prices slowly over time."

When you do this, you notice when you've accidentally built a portfolio of vibes instead of cash-flow machines.

### The two growth rates that matter: the business and your share

There's an investing nuance that separates adults from children: a business can grow while you, the shareholder, do not. What matters is *per-share* growth-earnings per share, cash flow per share, dividends per share. Your claim is measured per share.

If a company issues lots of new shares to fund acquisitions or pay employees, the business can get larger while your slice of it doesn't. Conversely, a company can have modest overall growth but strong per-share growth if it consistently buys back shares at sensible prices. That's one reason buybacks can be so powerful in mature businesses:

they convert stable cash generation into increasing ownership concentration for remaining shareholders.

Thinking per-share also protects you from a common mirage: "The company is growing!" is not enough. You want to know if your claim on the company's future cash flows is growing faster than the price you paid.

### When the engine misfires: reinvestment risk and capital allocation mistakes

Reinvestment is not automatically good. It's only good if management can reinvest at returns that exceed the company's cost of capital-plain language: if every dollar kept inside the business can reasonably become more than a dollar's worth of future value.

The two most common ways reinvestment goes wrong are painfully human. First, the empire-building instinct: executives love size, headlines, and "strategic transformations." Acquisitions can make sense, but they're also a socially acceptable way to overpay for growth. Second, the denial instinct: management keeps funding a project because admitting it failed is embarrassing. The money becomes "already spent," and humans have a hard time letting go of sunk costs.

From the outside, you can't control management. But you can choose what you own-and at what price. You can favor companies with a record of disciplined reinvestment and rational shareholder payouts. You can avoid stories that require perfection: heroic growth assumptions, permanently high margins, and flawless execution.

A wonderfully practical question to ask about any business is: "What would they do with an extra $10 billion?" The answer reveals a lot. Some companies would invest it in high-return projects. Some would pay it out. Some would light it on fire with expensive acquisitions. You don't need certainty; you just need to know which kind of management team you're betting on.

### The real reason diversified equities work: you're buying a portfolio of experiments

No single company's compounding is guaranteed. Competition is relentless. Technology rewrites business models. Consumer tastes shift. Regulation changes the playing field. The graveyard of once-great companies is long, and it's not done growing.

Diversified equity investing works precisely because you don't have to pick the immortal companies in advance. A broad index is a portfolio of experiments run by the global economy. Some businesses fail. Some stagnate. Some compound spectacularly. The index continuously reallocates weight toward the winners as they grow. You get exposure to progress without needing to predict which specific brand will dominate ten years from now.

This is also why "the stock market" is not the economy. The market is disproportionately a claim on the most profitable, scalable, and globally competitive businesses. When people say "stocks go up over time," they are really saying "a continually refreshed basket of leading businesses has tended to grow its earnings power over time." It's not a promise; it's an observed outcome of capitalism's selection process.

The emotional benefit of this framing is huge: you stop needing each holding to be a hero. You just need the system of ownership to do what it has often done—let the winners do more of the work.

### Practical translation: what ``owning the engine" should look like in your behavior

Owning equities responsibly means accepting that the quote can be wrong for long stretches. You're not paid for being right on Tuesday. You're paid for staying invested through discomfort while the underlying businesses keep compounding.

It also means being suspicious of narratives that turn ownership into entertainment. If your primary reason for holding a stock is a slogan, a charismatic CEO, a viral product demo, or a community's excitement, you're closer to fandom than investing. Real investing can still be fun, but it's fun in the way building something is fun—not in the way gambling is fun.

A grounded equity investor spends more time thinking about durability than drama: competitive advantages, pricing power, balance sheet resilience, and whether reinvestment has a high probability of being intelligent. Then they pair that with humility: diversification, low costs, and an acceptance that the best long-run plan often looks boring in the short run.

### Actionable takeaway

Pick one equity holding (or one equity fund) you already own and write a five-sentence "owner's memo" for it:

1. What does the business (or fund's companies) sell, in plain English?
2. Where does profit come from, and what could realistically threaten it?
3. How does management tend to use net income: dividends, buybacks, reinvestment, or acquisitions?
4. What would have to be true for this to compound well over the next 10 years?
5. What would have to go wrong for this to disappoint even if the economy is fine?

If you can't answer without leaning on the stock chart or a one-line narrative, reduce the position until you can-or replace it with a diversified equity vehicle whose compounding logic you *can* explain like an owner.

## 2.3 Decomposing Equity Returns

A stock chart is a thriller novel written in one-second chapters. It's all plot twists—gaps up, gaps down, "mystery selling," "risk-on," "risk-off." The problem is that your retirement doesn't care about plot. It cares about *physics*. And in equities, the physics is simple: in the long run, your return has to come from cash that shows up in your hands, cash that stays in the business and grows, and whatever the market decides to pay for those cash flows when you eventually sell.

That's it. There are only so many places return can hide.

This isn't a call to ignore prices. It's a call to stop treating price movement as an explanation. Price is an *output*. The explanation is the engine underneath and the discount the market applies to it.

### The three sources of equity return

A practical decomposition-often nicknamed the Gordon Growth Model in polite finance circles-breaks broad equity returns into three ingredients:

$$\textit{Total Return} \approx \textit{Shareholder Yield} + \textit{Fundamental Growth} + \textit{Valuation Change}$$

Each term has a personality.

*Shareholder yield* is the cash the market hands back to owners. Traditionally that meant dividends. In modern markets it also includes net buybacks (buybacks minus new shares issued). Dividends are the visible paycheck; buybacks are the quiet ownership-concentration scheme. Economically, both are ways the business transfers value to shareholders.

*Fundamental growth* is the increase in the business's per-share earning power over time-earnings per share, free cash flow per share, dividends per share. It's the compounding that happens because companies retain some profits and reinvest them, expand markets, improve margins, or simply keep pace with economic growth and inflation.

*Valuation change* is the mood swing. It's what happens when the market decides to pay a higher or lower multiple for the same stream of earnings-often discussed as P/E expansion or contraction. This is the speculative layer: it can dominate short horizons, then mean-revert with no regard for your feelings.

There's an almost therapeutic effect to naming these. Suddenly, "the market was down 18% this year" isn't a cosmic judgment. It's a statement that one or more of these ingredients went negative-usually the valuation piece-and you can start asking which part is likely to persist.

### A dinner bill analogy (because returns are a bill)

Imagine you buy a small restaurant.

- *Shareholder yield* is the cash you take home each year after paying expenses-money you can spend or reinvest elsewhere.
- *Fundamental growth* is the restaurant getting better: more tables, better reviews, higher prices, more repeat customers.
- *Valuation change* is what a stranger offers to pay you for the restaurant when you sell. Maybe they're euphoric about "the hot neighborhood." Maybe they're terrified of a recession. The offer can swing wildly even if the restaurant's Tuesday-night bookings haven't changed.

Over a year, the offer price can overwhelm everything else. Over twenty years, it's hard for the offer price to be the whole story, because the restaurant's cash generation either grew or it didn't. Markets can get excited, but they can't repeal arithmetic forever.

### Why dividends alone understate what shareholders actually receive

If you only count dividends, you miss a modern reality: many companies prefer buybacks to dividends because buybacks are flexible (no one panics if you "pause" a repurchase program the way they panic if you cut a dividend) and because they can be more tax-efficient for many investors.

So a more honest "cash returned to owners" term is *dividends + net buybacks*. Some practitioners call this *shareholder yield* or *augmented dividends*. Whatever name you choose, the point is to count *what the business is actually sending back to owners*, not just the part that arrives labeled "dividend."

There's a second reason this matters: buybacks can also sneak into the growth term. If a company earns the same total dollars but has fewer shares outstanding, earnings *per share* rise. That's real per-share growth for you as an owner-but if you're not careful, you'll count it twice: once as "growth" (because EPS went up) and again as "payout" (because buybacks happened). Clean decompositions require clean bookkeeping.

A simple rule keeps you honest:

> If your growth estimate is based on *per-share* metrics (EPS per share, cash flow per share), then your payout term should emphasize *dividends* and *net buybacks* as a cash-return measure, but you should avoid treating buybacks as an "extra" return on top of per-share growth *if your growth already includes the share-count effect.*

You can do it either way; you just can't do it both ways without accidentally inventing return out of thin air.

### The fundamental pair: yield + growth tends to be sticky

Over long spans, the "yield + growth" part behaves like a slow, heavy door. It doesn't slam around as easily.

Dividend yields don't jump from 1.5% to 8% overnight unless prices collapse. Earnings growth doesn't sustainably run at 25% forever for an entire economy unless something extraordinary is happening. These are anchored to economic reality: productivity, competition, inflation, demographics, innovation, and how much of corporate profits get reinvested versus paid out.

This is why long-term investors obsess over starting yields and plausible growth rates. Not because they're exciting, but because they're the part you can reason about without pretending to be clairvoyant.

If you're forced to forecast the next 20 years of broad market returns, the least embarrassing method is to:

1. Start with a plausible shareholder yield (dividends plus net buybacks, or dividends alone if you're being conservative).
2. Add a plausible real growth rate in per-share fundamentals.
3. Add expected inflation (because nominal earnings tend to rise with inflation over time).
4. Treat valuation change as *approximately zero* over very long horizons-or, if starting valuations are extreme, allow a modest headwind or tailwind.

That last part-valuation change-feels unsatisfying because it's the most dramatic. It's also the least controllable, and it's where most short-term forecasts quietly go to die.

### The speculative solo: valuation change dominates the short run

Over a year, markets can re-rate rapidly. A change in interest rates, risk appetite, or perceived uncertainty can make the market pay "18 times earnings" instead of "24 times earnings" without any immediate change in underlying earnings. That valuation move is real return (or loss) for you-if you're marking to market-but it's not the kind of return that compounds. It's a one-time repricing.

Valuation change is like the tide lifting or lowering a boat. You'll feel it. You can't reinvest it at the company level. And you can't count on it to repeat.

A subtle psychological hazard is that investors often confuse valuation gains with business success. When a stock doubles because the P/E doubled, it feels like the company "did great." In reality, the market simply became willing to pay more for the same earnings. That can reverse with brutal speed.

The reverse hazard is equally common: investors interpret valuation-driven drawdowns as "the business broke." Sometimes the business did break. Often it didn't. The market just changed the price tag on future cash flows.

### The stacked-bar reality: one year is a voting machine; twenty years is a weighing machine

The decomposition becomes vivid when you compare horizons.

Over one year, valuation change can easily be the biggest block. Fundamentals might add a few percentage points of growth and yield, while valuation swings add or subtract 20% like it's nothing.

Over twenty years, valuation change still matters, but it usually can't stay the main character. If the business (the collection of businesses in an index) doesn't grow and doesn't return cash, there is no sustainable mechanism for returns. Over long spans, fundamentals do the heavy lifting.

The following chart isn't "historical truth" for a specific period; it's a stylized illustration of a pattern investors keep relearning the hard way: short-term returns are often dominated by speculative rerating, while long-term returns are dominated by the fundamental pair-earnings growth and shareholder yield.

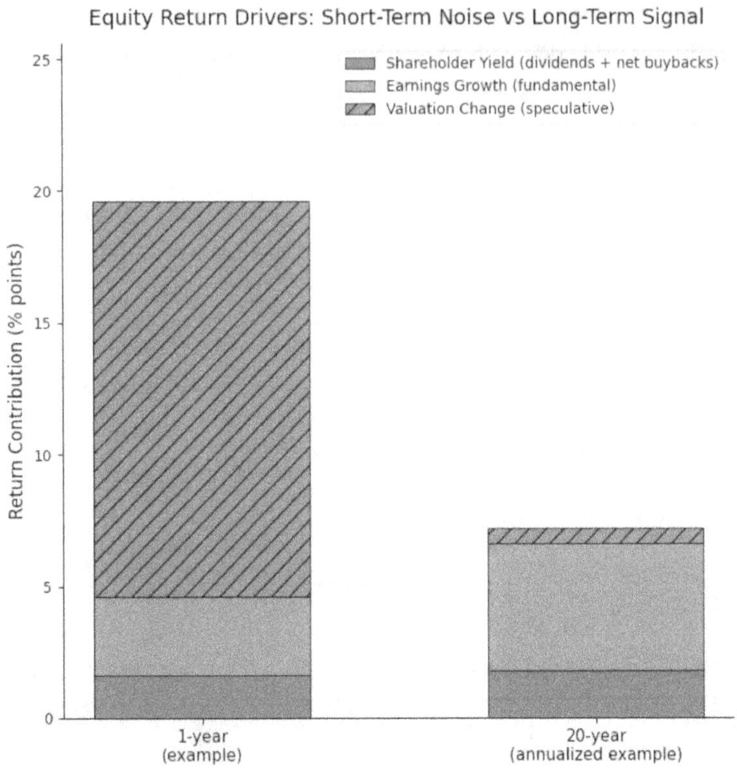


## Using the framework without fooling yourself

This decomposition is powerful precisely because it is incomplete. It doesn't claim to predict next quarter's return. It claims something more valuable: it forces you to explain where return *could plausibly come from*.

Here's how investors fool themselves with each term:

They overestimate *shareholder yield* by assuming buybacks are always additive. Buybacks at absurd prices can be value-destructive, like a company buying its own house at the peak of a housing bubble. The cash still goes out the door, but the value received per dollar is poor.

They overestimate *growth* by extrapolating great recent years into forever. Growth is competitive; competitors notice. Growth is cyclical; recessions happen. Growth is subject to gravity; huge companies can't grow at start-up rates without eventually becoming the economy.

They overestimate *valuation change* by treating mean reversion as a law that must pay them on schedule. Valuations can stay elevated or depressed longer than an investor's patience, career, or leverage can survive. Valuation is a timing tool with terrible customer service.

A more robust way to use the framework is to treat it as a range estimator rather than a point forecast. You can't know the exact path, but you can usually bound the future by refusing to assume miracles:

- Yield has a reasonable range given current market conditions.
- Long-run growth has a reasonable range given economic constraints.
- Valuation change is the wild card; assume zero as the baseline, then consider whether the starting point is unusually stretched.

If you do this, you gain something priceless: your expectations become harder to manipulate. When a talking head promises 15% annual returns "because innovation," you can calmly ask, "From which term? Show me the math."

### A concrete example of the "double-counting" trap

Suppose a company earns $10 billion and has 10 billion shares outstanding. Earnings per share are $1. If it uses cash to buy back 10% of shares, and next year earnings are still $10 billion, EPS becomes $1.11. Investors often celebrate this as "11% earnings growth."

But notice what happened: total earnings didn't grow. The per-share growth came from shrinking the share count-an ownership transfer funded by cash that might otherwise have been paid as a dividend.

That can still be great for shareholders. It just means you need to be consistent about whether you treat it as "growth" or "payout."

The discipline here isn't pedantry; it's self-defense. Most investor disappointment is not caused by bad news. It's caused by expectations that quietly assumed a free lunch.

### Why this matters for behavior, not just forecasting

When you understand the three drivers, you stop demanding that the market entertain you. You become less surprised by a "flat" decade that nonetheless delivered dividends and steady fundamental growth but suffered valuation compression. You also become less seduced by a roaring year that was mostly multiple expansion.

This is the emotional edge: the framework doesn't just tell you what returns are made of. It tells you what parts of your return are worth celebrating and what parts are worth treating with suspicion.

Fundamentals compounding is the kind of progress you can build a life on. Valuation expansion is dessert-enjoy it when it happens, but don't plan your nutrition around it.

### Actionable takeaway

Take the broad equity market you invest in (an index fund is perfect) and write down three numbers you can update once a year:

1. Current dividend yield (or shareholder yield if you track net buybacks).
2. A conservative long-run annual growth assumption for per-share fundamentals (pick a range, not a single point).
3. A valuation sanity check: is the market's multiple meaningfully above, below, or near its long-run norm?

Then create a simple expected-return range using:

$$\textit{Expected Return} \approx \textit{Yield} + \textit{Growth} \pm \textit{Valuation Change (small, uncertain)}$$

If the number you "need" for your plan is far above what this framework can justify, the solution usually isn't a better forecast. It's a better plan: save more, spend less, take more (intentional) equity risk, or extend the time horizon. The math gets a vote either way—better to invite it to the table early.

## 2.4 Defensive Assets: Bond Math and Rates

A bond is the least glamorous thing in finance: you lend money, you get paid back, you say thank you, and you go on with your life. Which is exactly why bonds are so easy to misunderstand. People look at their bond fund statement, see a negative year, and feel personally betrayed—like a seatbelt failed because the ride got bumpy.

The truth is more interesting. Bonds are not "safe" in the way a savings account feels safe. Bonds are safe in the way a contract is safe: you know the schedule of promised cash flows. What you *don't* know is what the market will charge you for that contract tomorrow, because tomorrow's interest rates may not match today's.

Once you see bonds as contracts that trade, you can finally understand the one bond fact that explains 90% of investor confusion:

> "When interest rates rise, existing bond prices fall. When interest rates fall, existing bond prices rise."

That inverse relationship isn't a quirk. It's a law of competitive pricing.

### Renting out your money

Think of lending like renting. If you rent out your car for $40/day and tomorrow everyone starts offering similar cars for $60/day, your

$40/day deal becomes less attractive. To rent it out, you'd have to lower the price of your car rental (or improve the car). Bonds work the same way: if new bonds are issued with higher yields, older bonds paying lower coupons must drop in price until their effective return is competitive.

This is why bonds can lose money even if the issuer is perfectly healthy. A bond can be a great credit risk and a terrible interest-rate bet at the same time. People often fuse these together—"safe borrower" equals "safe price"—and then get blindsided when rates move.

### The bond is a stream of cash flows, not a ticker

A plain-vanilla bond is a set of promised cash flows:

- periodic interest payments (coupons), and
- principal returned at maturity.

The market price of a bond is just the present value of those promised cash flows. "Present value" is finance-speak for "how much you'd pay today to receive money in the future, given the return you demand." If the demanded return goes up, the present value goes down. It's the same logic as valuing any long-lived asset—stocks included—except bonds are cleaner because the cash flows are (mostly) fixed.

Here's the key intuition: *the bond itself doesn't change when rates change.* The coupon checks are still the coupon checks. What changes is the *discount rate* investors use to value those checks. That discount rate is the market yield.

### Yield-to-maturity: the bond's ``internal rate of return,'' with fine print

Yield-to-maturity (YTM) is the single interest rate that makes the present value of a bond's promised cash flows equal to its current price. If you like plain English: it's the bond's implied annual return *if* you hold it to maturity, the issuer pays as promised, and you reinvest coupons at that same yield.

Those "ifs" matter.

Investors often treat YTM like a guaranteed outcome. It's not. It's a *pricing rate*—a way to summarize what the market is offering today. Your realized return will differ if you sell early, if rates change, if coupons are reinvested at different rates, or if credit conditions deteriorate.

Still, YTM is incredibly useful because it forces you to stop thinking in terms of "bond funds are down" and start thinking in terms of "what yield am I locking in today, and what risks am I taking to get it?"

### The teeter-totter: interest rates vs. bond price

The inverse relationship between rates and prices is easiest to hold in your head as a see-saw. When one side goes up, the other goes down. The only twist is the fulcrum—where the see-saw balances—represents *duration*.

Duration sounds like time, and it's measured in something like years, but what it really means is *sensitivity*. Duration estimates how much a bond's price will move for a 1% change in yields. Higher duration means bigger price swings for the same rate move.

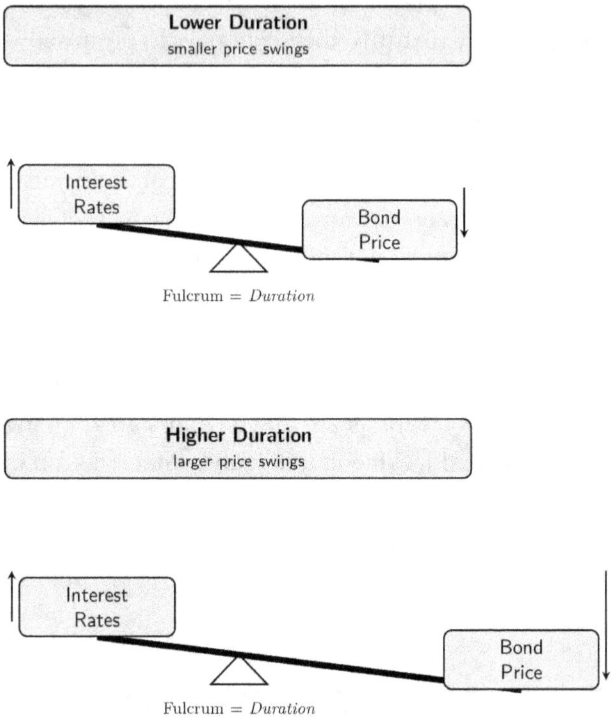


When market yields rise, bond prices fall. Duration tells you *how much* they fall for a given move in rates.

## Duration: the ``rate risk dial" you didn't know you were turning

If you own bonds through a fund, you might not know what you own in the only way that matters: how sensitive it is to interest rates. Duration is the dial.

A short-term Treasury fund has low duration. Rates can jump around and the price doesn't move much because you're getting your money back soon; the cash flows are close to today. A long-term Treasury fund has high duration. Its cash flows are far out in the future, so changing the discount rate changes the present value a lot.

A rough rule of thumb (and it really is a rough approximation) is:

$$\%\ Price\ Change \approx -Duration \times \Delta Yield$$

So if a bond has a duration of 7 and yields rise by 1% (one percentage point), the price might fall about 7%. If yields fall by 1%, the price might rise about 7%. Real bonds don't behave perfectly linearly—there's curvature in the relationship, called convexity—but duration gets you in the right neighborhood.

This is the source of a lot of "I thought bonds were safe" disappointment. People accidentally buy long duration because the yield looks a bit better, not realizing they just added a meaningful rate bet to the supposedly calm part of their portfolio.

### Why bonds can be a great defensive asset even though they sometimes lose money

Defensive doesn't mean "never negative." Defensive means "helps when your engines are sputtering."

High-quality bonds often play defense in two ways:

- They provide contractual cash flows you can spend or rebalance with.
- In many recession scares, yields fall as investors seek safety, which can push bond prices up—offsetting equity declines.

But defense is environment-dependent. If inflation rises and central banks tighten, yields can rise and bonds can drop at the same time stocks are struggling. That's not bonds "failing." That's you discovering which enemy you're fighting: recession fear or inflation fear. Bonds are a recession shield more than an inflation shield (unless they're inflation-linked, which is a different instrument with a different job).

The defensive role of bonds is also portfolio-specific. If your spending needs are near-term, bond cash flows are not just "diversification." They are scheduling. They let you fund life without having to sell equities at a bad moment. That's defense in the most practical sense: it reduces forced selling.

### The reinvestment problem: coupons are a feature and a risk

Bonds look simple because they pay interest. That interest is both your friend and a subtle risk. When a bond pays coupons, you'll likely reinvest them. The rate you earn on those reinvested coupons depends on the interest-rate environment at the time.

That means a bond investor faces *reinvestment risk*: if rates fall, your future coupons get reinvested at lower yields. Conversely, when rates rise, reinvestment becomes more attractive—but your existing bond's price declines in the meantime.

This is why bond math often feels like a set of trade-offs that refuse to let you be greedy in peace. Want more yield? You often take more duration (price sensitivity) or more credit risk (default sensitivity). Want less volatility? You usually accept lower yield. Bonds are a pricing machine for different kinds of safety.

### Convexity: the reason high-quality bonds sometimes surprise you (in a good way)

Duration is a straight-line approximation, but bond price behavior is curved. The curvature is called *convexity*. For plain, option-free bonds, convexity is usually positive, meaning: when yields fall, prices rise a bit *more* than duration predicts; when yields rise, prices fall a bit *less* than duration predicts (for the same size move).

This matters most for large rate moves and for longer-duration bonds. It's one reason high-quality long Treasuries can act like a shock absorber in deep risk-off events: if yields plunge, the price gains can be quite powerful.

There's a nasty exception: bonds with embedded options, like callable bonds (where the issuer can repay early). In those, your upside can get clipped when yields fall because the issuer wants to refinance away from you. That's "negative convexity," and it's why some bonds behave like they have a ceiling. Many investors discover this only after the fact, which is a painful way to learn geometry.

### The yield curve: why ``longer maturity" isn't just ``more return"

Investors often assume that lending longer should automatically pay more. Sometimes it does. Sometimes it doesn't.

Long-term yields reflect two forces:

- expectations for future short-term rates (where overnight and policy rates might go), and
- a term premium (extra compensation investors demand for locking money up longer).

That term premium can rise and fall for reasons that don't feel intuitive: changes in inflation uncertainty, risk appetite, supply and demand for safe long assets, and the market's appetite for duration risk. This is why long-term bond yields can move even when people's view of the near-term central bank path hasn't changed much.

For a bond investor, the practical point is humbling: buying long bonds is not just "getting a higher yield." It's taking a view—whether you admit it or not—on inflation uncertainty, growth risk, and the market's willingness to pay for long-term safety.

### Bond funds versus individual bonds: the difference that matters

Many investors feel safer owning individual bonds because there's a date on the calendar when they expect principal back. That can be true if you hold to maturity and there's no default. It creates a psychological anchor.

A bond fund doesn't mature. It continuously holds a ladder of bonds that roll down and get replaced. That makes the price fluctuate with rates, sometimes uncomfortably. But it also means the fund's yield resets over time as it buys new bonds at prevailing yields.

The key difference is not morality; it's *matching*. If you have a known future expense on a known date, individual bonds (or a defined-maturity bond ladder) can match that liability cleanly. If you want ongoing ballast in a portfolio, a fund can be an efficient, diversified instrument—as long as you understand the duration you're holding.

In both cases, the same rule applies: don't judge bonds primarily by last year's price change. Judge them by what yield you're earning going forward and what risk dial you've chosen to turn.

### How bond math protects you from ``yield envy"

The most common bond mistake is emotional: you see a higher yield somewhere else and assume it's a free upgrade. Bond math is the antidote because it forces the uncomfortable question: "Higher yield in exchange for what?"

Often the answer is "in exchange for duration," meaning your supposedly defensive asset will swing harder when rates change. Or "in exchange for credit risk," meaning the bond might behave more like an equity substitute during downturns. Or "in exchange for optionality," meaning your upside is capped when times are good and your downside remains when times are bad.

Once you internalize that trade, the bond market stops feeling like a list of coupons and starts feeling like a menu of risks with prices attached. That's a much healthier relationship.

### Actionable takeaway

Look up one number for every bond holding you have (funds included): *duration*. Then do a simple stress test on paper:

> "If yields rise by 1% tomorrow, approximately how much might this position fall?"
> Answer: $\approx$ *duration* $\times$ 1%.

If the answer makes you flinch, you're not holding a "bad" bond—you're holding a bond whose job description doesn't match your nerves or your timeline. Adjust the dial deliberately: shorter duration for stability and near-term spending needs; longer duration only if you consciously want rate sensitivity as part of your defense.

## 2.5 Credit Risk and the Search for Yield

The phrase "bond investor" sounds like someone who dislikes excitement. Then you meet the modern bond investor, staring at a screen that says *4.9%* on Treasuries and *7.8%* on high-yield corporates, and you can practically hear the internal bargaining: "It's still a bond. How bad can it be?"

That thought—*it's still a bond*—is the doorway to a very expensive misunderstanding. Bonds are not one asset class. They're a family of contracts with different enemies. Government bonds mostly fight *interest-rate risk*. Corporate bonds fight interest-rate risk *and* the possibility that the borrower becomes a different creature under stress: weaker, desperate, or insolvent. High-yield bonds go a step further and introduce a truth that polite asset allocation charts often hide: a "bond" can behave like a stock when you need it most to behave like a seatbelt.

Credit risk is the reason.

## Two kinds of ``risk-free": price risk and promise risk

When people say U.S. Treasuries are "risk-free," they usually mean "risk-free of default in nominal dollars." In plain English: the U.S. government can tax and issue currency, so a Treasury bond is widely treated as a promise that will be paid (in dollars) as scheduled. That doesn't mean Treasury prices can't fall; they can fall dramatically when yields rise. It means the promise—the contractual cash flows in nominal terms—is treated as exceptionally reliable.

Corporate bonds are different. They are promises from entities that can't print money. Companies can and do fail. Even when they don't fail, they can become impaired: profits evaporate, refinancing becomes costly, and what looked like a boring interest payment starts to feel like a negotiation with reality.

So "risk-free" depends on what you mean by risk:

- Treasuries have plenty of *price risk* (especially at longer durations), but very low *promise risk* in nominal terms.
- Corporate bonds have both *price risk* and *promise risk*.

If you're building a portfolio that needs true defensive behavior—assets that can fund spending, allow rebalancing, and provide psychological ballast—promise risk is not a rounding error. It's the whole plot.

## Credit spreads: the market's price tag on ``might not get paid"

Credit risk shows up in one number that quietly runs the bond universe: the *credit spread*. A spread is simply the extra yield a corporate bond offers above a comparable Treasury. That extra yield is the market's compensation for taking corporate-specific risk: default risk, downgrade risk, liquidity risk (harder to sell in a panic), and the general unpleasantness of owning promises from borrowers who can get into trouble.

If a Treasury yields 4% and a corporate bond yields 5.2%, the spread is 1.2 percentage points. That 1.2% is the bond market saying: "This borrower is not the government. Pay me."

Spreads are not fixed. They breathe with fear and confidence. When the economy is calm and investors are hungry for income, spreads compress—meaning the extra compensation for risk shrinks. When recession fears rise, spreads widen, often violently, because investors suddenly care a lot about who might not be able to refinance their debt.

That breathing is the reason credit risk can wreck the "defensive" role of bonds. In a bad risk-off event, spreads can widen at the same time equities fall. If you were counting on corporate bonds to be a stabilizer, you discover you bought something closer to an equity cousin than a Treasury sibling.

### Investment grade vs. high yield: the line that matters in a crisis

The corporate bond market draws a bright psychological line between *investment-grade* and *high-yield* (often called "junk"). Investment-grade issuers are considered relatively strong borrowers. High-yield issuers are, by definition, weaker credits—more leverage, more cyclicality, more fragility, or some combination.

What matters isn't the insult in the nickname. What matters is how these bonds behave when the lights flicker.

High-yield bonds tend to be most vulnerable precisely when diversified portfolios need defense: recessions, liquidity squeezes, and moments when refinancing becomes difficult. In those periods, investors don't ask "What's the coupon?" They ask "Do I get my money back?" That question isn't theoretical. It changes prices fast.

A useful way to describe the difference is this: investment-grade bonds are usually priced as "loans to businesses." High-yield bonds are often priced as "a loan *plus* an option on survival." That survival option is what explodes in value (or collapses) when the economy turns.

## Why ``higher yield" is often just ``hidden leverage"

It feels natural to crave yield. Yield is immediate, concrete, and easy to compare. A 7% yield seems objectively better than a 4% yield in the same way that a larger paycheck seems better than a smaller one.

The catch is that yield is not a free lunch; it's a receipt for risk.

When you "reach for yield," you usually accept one (or more) of these hidden trades:

- *More credit risk*: the borrower is weaker.
- *More liquidity risk*: the bond will be harder to sell at a fair price under stress.
- *More cyclical exposure*: the borrower's cash flows depend on a healthy economy.
- *More leverage sensitivity*: small changes in business conditions have outsized effects on the ability to service debt.

Those trades have a pattern: they tend to hurt at the same time. When the economy slows, liquidity dries up, defaults rise, and risk appetites vanish together. That co-movement is why high-yield bonds can fail as diversifiers. They don't just have risk—they have *timing* risk, and the timing is rude.

If you want a blunt translation: high yield is often "equity-like downside with bond-like upside." You earn a higher coupon, but in exchange you accept that bad outcomes can look a lot like stock drawdowns without the long-run compounding engine that makes equity risk worthwhile.

### The ``bond wrapper'' illusion

A huge fraction of investing mistakes come from believing the wrapper instead of the contents. People see the word "bond" and assume "anchor." But an asset class label does not guarantee economic function.

A high-yield bond fund can have daily price volatility that surprises investors. More importantly, its drawdowns tend to appear right when equity investors are already stressed. The pain is not just the loss. The pain is that the thing you bought for comfort shows up to the crisis wearing the wrong uniform.

This is where portfolio function matters more than category. If the job is "reduce portfolio crash risk," you want assets that tend to do well (or at least hold up) when risk assets are falling. If the job is "generate income," high-yield may help. If the job is "income *and* defense," high-yield is often a trap: you get the income until the day you most need the defense, and then the income doesn't compensate you for the mark-to-market damage.

### Default is not the only way credit hurts you

Investors hear "credit risk" and picture bankruptcy headlines. Default is the dramatic version, but the market punishes credit in quieter ways too.

One is *downgrade risk*. A company can remain solvent but be downgraded, which typically means its bonds must offer higher yields to attract buyers. Higher yields imply lower prices. You can lose money without any missed payments, simply because the market reclassified the borrower as more fragile.

Another is *refinancing risk*. Many companies don't "pay down" debt in a steady march; they roll it over. When conditions are easy, this looks harmless. When conditions tighten, the cost of rolling debt rises, and the business can go from "fine" to "problem" in an alarmingly short time.

And then there's *liquidity risk*. In calm markets, you can sell almost anything at a reasonable price. In stressed markets, the bid disappears, and the bond market reveals its structure: many corporate issues are not built for instant exits at fair prices. Liquidity is not something you own; it's something the market *lends* you—and it can call the loan.

### The cycle: when credit feels safest is often when it's least rewarding

Credit spreads tend to be narrow when the world feels safe. That's also when you get paid the least to take credit risk.

This is the part that feels unfair: the best time to buy credit (purely from a compensation standpoint) is often when you least want to—when spreads are wide and headlines are ugly. The worst time to buy credit is often when it feels like free money—when spreads are tight, defaults are low, and everyone is convinced the soft landing is eternal.

This doesn't mean you should try to time credit like a trader. It means you should have an emotional warning label: when yield looks "easy," it's probably priced that way because many investors are reaching for it at once.

### A practical sorting hat: what kind of bond is this, really?

If you're holding corporate bonds (directly or through funds), one question clarifies almost everything:

> If equities drop 30% in a recession scare, do I expect this bond holding to be a helper, a bystander, or a co-sufferer?

High-quality Treasuries have a decent chance of helping (depending on the inflation backdrop). Short-term government cash equivalents are usually bystanders (stable, not heroic). High-yield credit is often a co-sufferer.

That sorting doesn't require precise forecasts. It requires honesty about the economic linkage. A highly leveraged company's ability to pay is tied to business conditions and financing conditions—the same conditions that often hurt equities. You don't diversify equity risk by buying a different wrapper around a similar dependency.

### Where corporate bonds *do* make sense

Credit risk is not evil. It's a tool. It can be appropriate when the role is clearly defined.

Corporate bonds can make sense as a middle layer for investors who want income and are comfortable taking some economic risk, especially in high-quality investment-grade credit. They can also make sense as part of a broader fixed-income allocation when the investor understands that "fixed income" includes multiple sub-jobs: liquidity, stability, and income.

High-yield bonds can make sense too—particularly when priced attractively and sized appropriately—as a return-seeking asset that you treat honestly: not as a crash protector, but as a risk asset with a coupon. The mistake is not owning high yield; the mistake is *believing* it will behave like an anchor.

### The behavioral trap: income as emotional anesthesia

Yield has a special psychological power because it feels like progress even when the account value isn't growing. A monthly distribution can make a risky asset feel safer than it is—like a noisy engine that sounds healthy because it's loud.

This is one reason "income investing" can become an identity rather than a strategy. People start defending yield the way they defend a lifestyle. They tolerate hidden risks to preserve the feeling of being paid.

A clean way to break the spell is to force a total-return view:

> A 7% yield with a 15% drawdown is not "income." It's a return profile.

The market doesn't care which part of the return you emotionally prefer. Your future spending power only knows the total.

### Actionable takeaway

For every bond fund or bond allocation you own, write down two numbers and one sentence:

1. *Credit quality*: roughly how much is Treasury/agency, investment-grade, and high-yield (fund fact sheets disclose this).
2. *Duration*: your interest-rate sensitivity (how much price can move when yields change).
3. *Job description*: "This holding is here to ________________ during equity drawdowns."

If you can't write a sentence that you genuinely believe under stress, resize or replace the position until you can. The point isn't to avoid credit risk forever. The point is to stop accidentally buying equity-like drawdowns in the part of your portfolio you were relying on for sleep.

## 2.6 Real Assets and Inflation Protection

Inflation is the only financial risk that pretends not to be a risk. It doesn't arrive as a red number on your brokerage screen. It arrives as a quiet betrayal: the same life costs more while your "safe" dollars buy less. You can go years without noticing it, then suddenly realize you've been losing ground the whole time—like walking up an airport moving walkway that's going the other direction.

That's why inflation protection is psychologically tricky. A 25% stock drawdown feels like a crisis. A 25% loss of purchasing power spread over a decade feels like "just the way things are." But for long-duration wealth, inflation is not background noise. It's the tax you pay for being paid in nominal dollars.

Real assets exist for one reason: they're closer to the physical economy than a financial promise. They're not just pieces of paper (or pixels) that say you're owed money. They are claims on things people use: shelter, energy, metals, farmland, infrastructure. When the unit of account (the currency) changes in value, the prices of useful things often adjust—sometimes quickly, sometimes painfully, often in a way that surprises investors who built their portfolios as if a dollar were a constant.

## Two kinds of inflation pain: expected and unexpected

Most markets can live with *expected* inflation. If everyone anticipates 2% inflation, contracts, wages, and interest rates gradually adapt. The real damage comes from *unexpected* inflation—when prices jump faster than the system's ability to renegotiate.

Unexpected inflation is where traditional anchors can feel like they've betrayed you. A nominal bond promising fixed payments becomes less valuable in real terms. Cash becomes a wasting asset. Even equities can wobble if higher inflation pushes interest rates up and compresses valuation multiples.

The goal of inflation protection is not to "beat inflation every year." The goal is to reduce the chance that a decade of rising prices quietly erodes the lifestyle your portfolio was built to fund. In other words: you want at least some of your wealth tied to things that can reprice when the world reprices.

### TIPS: the rare hedge that is not a vibe

Treasury Inflation-Protected Securities (TIPS) are unusual in finance because they do exactly what they say on the label.

A TIPS bond is a U.S. Treasury bond whose principal adjusts with inflation as measured by the Consumer Price Index (CPI). The coupon rate is fixed, but it's applied to the inflation-adjusted principal, so the dollar amount of interest you receive rises with inflation. At maturity, you get the greater of the inflation-adjusted principal or the original principal amount—so there's a floor at maturity against deflation wiping out your principal.

That structure is an explicit inflation linkage. It's not an "inflation-sensitive business model" story. It's not "this asset tended to do well in the 1970s." It's a contract that says, in effect, "We will adjust what you get paid based on CPI."

There are still risks. TIPS can fluctuate in price if real interest rates change. They can be unpleasant in a rising real-rate environment even if inflation is high. And CPI is a particular basket, not a perfect match for any individual household's expenses. But as an instrument, TIPS are about as close as markets offer to a clean, direct hedge of inflation in high-quality fixed income.

A practical way to think about TIPS is: nominal Treasuries protect you from *deflation and recession scares* more cleanly; TIPS protect you from *inflation surprises* more cleanly. They are cousins, not duplicates.

### CPI isn't "your" inflation-and that's not a dealbreaker

CPI is a weighted index based on what consumers buy, averaged across a population. Your life is not an average. If you spend heavily on healthcare, childcare, rent, or tuition, your inflation may run hotter than CPI. If you own your home outright and your spending leans toward goods whose prices fall over time, your inflation may run cooler.

This mismatch can tempt people to dismiss inflation hedges as pointless: "If it doesn't match me perfectly, why bother?" That's like refusing to wear a seatbelt because it doesn't prevent every injury. The point is to reduce exposure to a broad, systemic risk—currency purchasing-power erosion—using a transparent reference standard.

A good inflation plan accepts imperfection and aims for robustness. You won't hedge every expense category. You can hedge the risk that the currency unit itself is quietly shrinking.

## Real estate: inflation linkage with a thousand footnotes

Real estate is the most popular "inflation hedge" people already own—often unintentionally—because their home is both a place to live and a real asset. But real estate as an inflation protector depends on *what kind* of real estate and *how* you own it.

If you own a home with a fixed-rate mortgage, you have a powerful built-in hedge: your monthly payment is fixed in nominal dollars while wages and prices tend to rise over time. Inflation can actually *help* a fixed-rate borrower because the debt gets easier to pay back in real terms. The house itself may rise in nominal value, not because it's a magic asset, but because replacement costs and land scarcity can rise with the price level.

Rental real estate introduces another channel: rent. In many markets, rents can adjust with inflation, especially in shorter-lease residential settings. That makes rental property feel like a "cash-flowing inflation hedge."

And then the footnotes arrive.

Property taxes can rise. Insurance can rise. Maintenance costs can rise. Local supply can change. Regulations can cap rent increases. Vacancy risk is real. Real estate is not just inflation sensitivity; it's a full-contact sport with idiosyncratic risks. It can be an excellent real asset, but it's not a free hedge.

Public real estate through REITs (real estate investment trusts) adds another layer. REITs are designed to distribute a high share of taxable income to shareholders, which often translates into meaningful dividend yields. That can make them feel like a hybrid between an equity engine and an inflation-linked income source.

But REITs are still equities. Their prices can swing with interest rates and risk appetite. In sharp rate rises, REITs can fall even if long-run property cash flows remain viable, because the market reprices the discount rate applied to those cash flows. A REIT can be a real asset in terms of underlying holdings and still trade like a financial asset in terms of day-to-day behavior.

If you want to use real estate as a shield, the cleanest framing is: real estate has *some* inflation pass-through over time, but it can be very noisy over shorter horizons and is sensitive to financing conditions.

### Commodities: the inflation hedge that doesn't care about your calendar

Commodities are the raw inputs of inflation headlines: energy, metals, agricultural goods. When commodity prices surge, inflation often follows—not always, but often enough that commodities have earned a reputation as an inflation-sensitive asset class.

The intuition is straightforward: if oil, copper, wheat, and natural gas get more expensive, the physical economy pays the bill. That bill shows up in transportation, manufacturing, food, and heating costs. Commodities are closer to the source.

But commodities are also the inflation hedge most likely to hurt your feelings. They can be volatile. They can go through long stretches of mediocre returns. They don't generate cash flow like a business does or contractual payments like a bond does. They are prices—driven by supply disruptions, geopolitics, weather, technological changes, and global demand.

There's also a mechanical nuance for investors who access commodities through futures-based funds: your return comes from multiple components, not just spot price changes. Futures can trade above or below spot, and contracts need to be rolled, which can create a drag or a boost depending on the shape of the futures curve. This is one of those areas where "real asset" doesn't mean "simple."

So why bother? Because commodities can be one of the few things that respond quickly to inflation shocks. They are often less about compounding and more about insurance. Like many forms of insurance, they can feel annoying for years and suddenly feel genius in a year you didn't want to live through.

### Gold: a hedge against monetary stories

Gold is the asset people reach for when they stop trusting promises.

It's a peculiar kind of real asset: it's tangible, it's scarce in a way humans find psychologically satisfying, and it has thousands of years of narrative weight. But it doesn't produce income. It doesn't pay coupons. Its expected real return over very long horizons is debated, and its short- and medium-term behavior can be driven by real interest rates, currency moves, and shifting investor sentiment.

So what is gold *for*?

A practical answer is: gold is a hedge against certain tail risks—currency debasement fears, geopolitical stress, and episodes where confidence in financial assets wobbles. It can act as a portfolio diversifier in some regimes, but it's not a precise CPI hedge in the way TIPS are.

Gold is also a behavioral tool. Some investors hold it not because they can model it perfectly, but because it gives them the courage to hold their engines through ugly times. If a small allocation helps you avoid a catastrophic behavioral mistake, it can earn its keep even if its long-run expected return is not thrilling.

The mistake is turning gold into a religion. The point of a shield is to reduce a specific vulnerability, not to become the centerpiece of your identity.

### Stocks as partial inflation hedges: long-run truth, short-run frustration

Businesses live in the real economy. They sell real things and pay real wages. Over time, many companies can raise prices, adapt, innovate, and maintain real earnings power. That makes diversified equities a partial inflation hedge over long horizons—especially compared to cash.

But the short-run can be the opposite of comforting. Inflation often brings higher interest rates, and higher rates can compress valuation multiples. In those regimes, stock prices can struggle even if nominal earnings rise. Investors experience the whiplash: "Inflation is high, companies can raise prices, so why are stocks down?" Because the market is repricing the discount rate and the risk premium, not just this year's revenue.

This is why using equities as your only inflation plan is risky. They may protect purchasing power over decades, but they are not designed to be a clean, targeted hedge for unexpected inflation spikes.

### Choosing shields the same way you choose anchors: by job description

Inflation protection only works when you define what "protection" means for you.

If the problem is "I need safe, predictable purchasing power for expenses 5–15 years from now," TIPS start to look like a purpose-built tool. If the problem is "I'm worried about a burst of unexpected inflation," a modest allocation to real assets (including commodities or commodity-linked equities, depending on implementation) can help.

If the problem is "I'm nervous about long-run monetary credibility," gold might be a psychological stabilizer—again, in modest size. If the problem is "I want inflation-sensitive income streams," some forms of real estate may help, but you should treat them as hybrid assets with rate sensitivity, not as pure shields.

The deeper point is the same one that keeps showing up across asset classes: label is not function. The job is function.

### The hidden cost of shields: they're often inconvenient on purpose

Great inflation hedges tend to be awkward. Real assets can be illiquid, volatile, or tax-inefficient. TIPS can be misunderstood and can have uncomfortable price moves when real rates rise. Commodities can feel like dead weight until they suddenly don't. Real estate can call you at 2 a.m. with a burst pipe.

That inconvenience is not an accident. It's the price of diversification. If inflation protection were smooth, obvious, and always rewarding, everyone would own it heavily, and it would stop being cheap insurance.

The goal isn't to build a portfolio that never has awkward parts. The goal is to build a portfolio where the awkward parts pay you back during the specific scenario that would otherwise wreck your plan.

### Actionable takeaway

Write down your personal inflation exposure in one sentence, then match it to an instrument:

> "My biggest inflation vulnerability is ______________________________."

Then do a simple portfolio audit:

- If your vulnerability is *near- to mid-term spending power*, consider whether you have any explicit inflation-linked assets (like TIPS) sized to matter.

- If your vulnerability is *unexpected inflation shocks*, consider whether you have a modest real-asset sleeve (commodities and/or real estate exposure) that you can hold through ugly volatility.
- If your vulnerability is *behavior under stress*, consider whether a small "belief hedge" (often gold, sometimes something else) would help you stick with your long-run engine without panicking.

Inflation is slow, but it is not gentle. The most practical defense is not a perfect hedge—it's a deliberate one.

## Chapter 3

# Portfolio Architecture

*If picking stocks is like choosing furniture, portfolio architecture is building the house itself. You wouldn't buy a couch before pouring the foundation, yet many investors obsess over individual tickers while ignoring the structural integrity of their wealth. In this chapter, we move from analyzing parts to designing the whole. We will learn how to engineer a portfolio that doesn't just chase returns, but mechanically suppresses the risks that trigger our worst behavioral instincts.*

## 3.1 Defining Pain: Volatility vs. Drawdown

A portfolio statement has a peculiar talent: it can turn an adult into a superstitious gambler in under thirty seconds.

You open the app, expecting a calm number—a tidy summary of your competence. Instead, you get a red percentage that feels like accusation. Your brain doesn't translate it as "my assets are temporarily marked down." It translates it as "I did something wrong" or, worse, "I'm about to be poor." That is risk in the only sense that matters: the sensation that makes you flinch, panic, bargain, freeze, or smash the sell button.

Finance likes to define risk as *volatility*: how much returns bounce around. It's clean, measurable, mathematically elegant, and emotionally tone-deaf. Volatility counts good surprises and bad surprises as the same kind of "danger," the way a heart monitor would treat excitement and terror as identical because both spike your pulse. The market, meanwhile, does not ruin people with excitement. It ruins them with despair at exactly the wrong moment.

So here's a better definition, one that fits both the math *and* the mirror: volatility is the *price of admission*. Drawdown is the *risk of ruin*.

## The price of admission: the market's cover charge

If you want the long-run rewards of risky assets—ownership in businesses, claims on future cash flows, participation in economic growth—you pay a cover charge. That cover charge is the daily, weekly, and yearly wobble of prices. You do not get to negotiate it away. You can't "diversify out" the fact that markets move. You can only decide how much movement you can tolerate without sabotaging yourself.

Volatility is the market reminding you that the future is not politely predictable. It is the constant background noise of uncertainty. And it's not just noise in a chart: it's noise in your head. It shows up as the temptation to check your portfolio too often, to reinterpret every headline as a personal signal, to confuse motion with meaning.

Yet volatility, by itself, is not the villain investors think it is. Most volatility is *recoverable* discomfort. A portfolio can be volatile and still be fundamentally fine. The problem is that your nervous system doesn't grade on a curve. It experiences volatility as a series of threats. The more often you feel threatened, the more likely you are to do something irreversible.

This is why "risk tolerance" questionnaires often fail in real life. They ask you how you feel about hypothetical numbers. Your body, in a real drawdown, answers a different question: *how much pain can you endure before you need the pain to stop?* Markets don't care whether your stopping point is rational.

## Risk of ruin: the drop that breaks the compounding engine

Drawdown is a different creature. Drawdown measures the fall from a previous peak. It's what you feel when you say, "I was up, and now I'm not." It is the psychological punch of lost progress.

There are two ways drawdown hurts you, and one of them is pure arithmetic.

The arithmetic is brutal because losses are not symmetric. A 10% loss is not "canceled" by a 10% gain. Once you lose money, you have a smaller base to grow from. That is the part of investing that feels unfair because it *is* unfair—mathematically, not morally.

If your portfolio drops 50%, you need a 100% gain to get back to even. Not to get rich. Not to beat inflation. Just to return to your starting point, like a swimmer who's been pulled out by a rip current and now has to fight twice as hard merely to touch the same beach.

That asymmetry is why deep drawdowns are the compounding engine's enemy. Compounding thrives on time and consistency. Deep drawdowns steal both: they steal the base, and they steal the behavior. Investors don't merely suffer a 50% drawdown; many investors *change their plan* at the bottom. They convert a temporary mark-to-market loss into a permanent, locked-in loss.

### The ``Math of Loss": why big holes are hard to climb out of

The recovery math is simple enough to do on a napkin, and sharp enough to permanently change how you think about "risk."

Suppose you start with $100.

- Lose 10%: you're at $90. To get back to $100, you need $10 of gains on a $90 base: about 11.1%.
- Lose 25%: you're at $75. To get back to $100, you need $25 on a $75 base: 33.3%.
- Lose 50%: you're at $50. To get back to $100, you need $50 on a $50 base: 100%.
- Lose 75%: you're at $25. To get back to $100, you need $75 on a $25 base: 300%.

That curve is not linear. It gets steeper the deeper you go, like debt compounding against you. And the market does not offer a special "recovery mode" after you've been crushed. The expected return doesn't double just because you need it to.

Near the discussion of recovery math, the non-linear curve deserves to be seen, not merely understood.

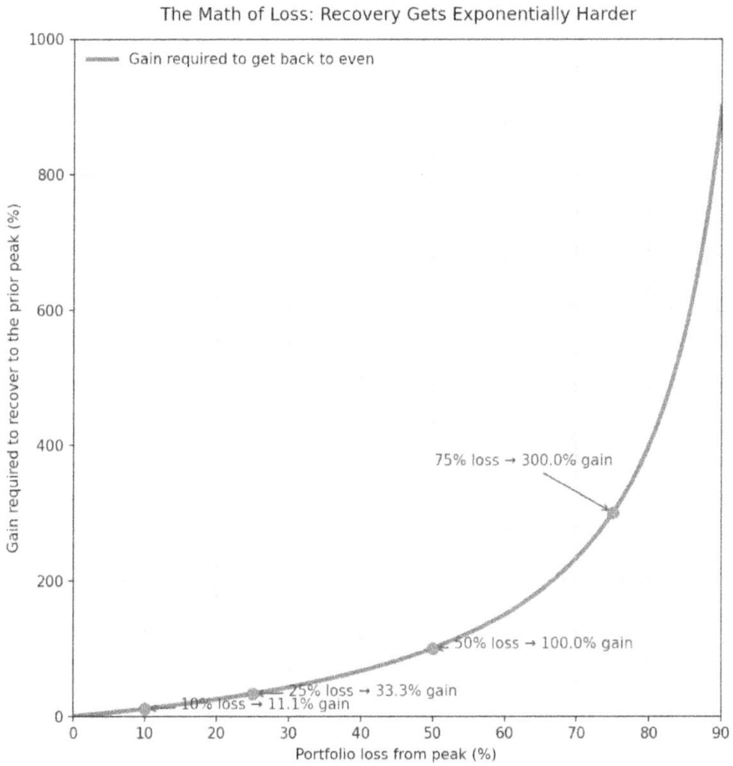


Once you've internalized that picture, a lot of investing advice starts to sound different. "Just hold on" becomes less like a motivational poster and more like a strict engineering requirement. "Avoid catastrophic mistakes" stops being conservative and starts being logical. You can chase upside for a decade and lose the whole game with one unmanageable drawdown—not because you were wrong about the future, but because you were wrong about yourself.

### Two investors, same average return, wildly different lives

Here's a trap investors fall into because we are trained, from school onward, to average things.

Imagine two portfolios that both average roughly 8% per year over a long horizon. Portfolio A is a smooth-ish climber: it has setbacks, but they're modest. Portfolio B is a thrill ride: it rockets up, then falls off cliffs, then recovers, then does it again.

If you only look at the average, you might conclude these are equivalent. Your stomach disagrees. Your behavior disagrees. And the math, quietly, often disagrees too.

Why? Because the *sequence* matters. When big losses show up, they change what future gains can do. Even if Portfolio B eventually claws back, it spent time digging out of holes instead of building new ground. And during those holes, investors tend to "help" by selling, pausing contributions, or switching strategies. They don't do this because they're stupid. They do it because they are in pain, and pain demands relief.

The market does not send you a calendar invite saying, "This drawdown is temporary; please remain calm." It sends you a feeling: *this might never end*. That feeling is the behavioral edge of drawdown. It's where plans break.

### Why volatility gets blamed for what drawdown did

Volatility is visible. You see it daily. It gives you something to point at. Drawdown, meanwhile, feels like a story about your competence: "I was up 30%, now I'm only up 5%. I blew it." That story is more personal, so investors often disguise it as a technical complaint: "This is too volatile."

But the specific kind of volatility that wrecks people is volatility that shows up as *a deep, sustained drop from a peak*. That's drawdown. And the peak matters. Humans anchor to recent highs the way hikers anchor to the summit they can see from the trail. You don't measure your progress from sea level; you measure it from the last ridge you reached.

That's why bull markets create fragile investors. When you've only known upward motion, the first serious drawdown feels like an emergency. You're not merely losing money; you're losing the identity of being the kind of person who makes money. That's a hard identity to surrender gracefully.

### Recovery is not just a number; it's a time commitment

The chart shows a percentage, but the real hidden cost is *time*.

A 50% drawdown requires a 100% gain. Even if your long-run return expectation is, say, 8%—10% per year, a 100% recovery is not a quick errand. It's years. During those years, you aren't "making progress" in the way you imagined. You are restoring what you already had. You are rebuilding the runway before you can take off again.

Time is the scarcest compounding input. You can contribute more, cut fees, optimize taxes, and tweak allocations, but you cannot manufacture more decades. Deep drawdowns steal years because they push your portfolio into a state where it must first heal before it can grow.

This is why investors nearing a spending goal—retirement, a down payment, college tuition—experience drawdown as existential. The portfolio may recover eventually, but *your timeline might not*. A drawdown right before you need the money is not the same as a drawdown when you're 27 and still contributing. Same market. Different stakes.

**Volatility is two-sided. Drawdown is one-sided. Your heart knows which side you live on.**

Volatility treats a +5% month and a -5% month as equal magnitude. From a statistical perspective, that symmetry is tidy. From a human perspective, it's absurd.

Gains rarely cause forced selling. Losses do. Gains rarely trigger career risk for professional managers. Losses do. Gains rarely cause sleep deprivation. Losses do. In short: volatility is the measurement a spreadsheet prefers. Drawdown is the measurement your life experiences.

Drawdown is also path-dependent: the order of returns matters. A portfolio that goes up 30%, down 20%, up 10% is not the same experience as one that goes down 20%, up 30%, up 10%, even if the ending number is close. The first one creates the sensation of giving back winnings. The second creates the sensation of recovery and competence. Same math. Different psychology. Different likelihood of sticking with the plan.

A good risk definition has to bridge both worlds. If it can't predict behavior, it can't predict outcomes, because outcomes depend on behavior.

### The hidden link between drawdown and long-term return: the ``behavioral gap"

Most investors don't fail because they never heard the right advice. They fail because they *could not execute* the right advice under stress.

A deep drawdown is stress. Under stress, humans narrow their attention, shorten their time horizon, and crave certainty. The market, during a drawdown, offers none. That mismatch creates a predictable pattern:

- You demand a reason.
- You find a narrative (usually terrifying, sometimes righteous).

- You act to relieve discomfort.

The relief is real. The cost shows up later.

This is the core tragedy of compounding: the biggest damage often comes from decisions made in a small number of emotionally extreme weeks. The investor who sticks with the plan doesn't win because they forecasted better. They win because they avoided the irreversible mistake.

That's why deep drawdown is better described as "risk of ruin" than as "high volatility." It's not the wiggles that kill you. It's the wiggles that become a cliff, plus the very human instinct to jump off it to stop the falling sensation.

## Practical definitions you can actually use

It helps to keep two different questions in your head, because they lead to different portfolio choices.

*Volatility question:* "How much up-and-down movement can I live with and still behave like a grown-up?"

*Drawdown question:* "What is the worst peak-to-trough loss I can tolerate without breaking the plan or jeopardizing real-life obligations?"

The first question is about comfort. The second is about survival. Comfort matters because discomfort leads to mistakes. Survival matters because survival keeps the game going long enough for compounding to matter.

Investors often get stuck because they answer the first question with bravado and the second question with denial. They say they can handle volatility—until volatility shows up as a drawdown that threatens their timeline, their identity, or their sense of safety.

### A quick way to stress-test your own ``sleep-at-night" limit

Pretend your portfolio hits a new all-time high on Monday. On Friday, it's down 30% from that peak.

Nothing about your job has changed. Nothing about your family has changed. The only change is the number.

Now ask yourself, without trying to sound tough: what do you do next?

If your honest answer includes "sell some to be safe," "wait until things calm down," or "move to cash for now," then a 30% drawdown is beyond your behavioral limit. That doesn't make you weak. It makes you human. It means your portfolio is asking you to perform a behavior you are not wired to perform.

The goal isn't to become a hero. The goal is to design a portfolio that makes heroism unnecessary.

#### Actionable takeaway

Write down one number: the maximum peak-to-trough drawdown you believe you can endure *and still follow your plan* (not the number that makes you sound brave). Then translate it into a portfolio rule: if your current allocation has historically produced drawdowns larger than that number, reduce risk *now*, while you're calm. The best time to choose your pain is when you're not already in it.

## 3.2 The Free Lunch: Diversification and Correlation

Picture two street vendors who share a wall. One sells ice cream. The other sells umbrellas.

On a bright, hot Saturday, the ice-cream line curls around the corner while the umbrella guy stares at a rack of untouched inventory, basically running a small museum of regret. On a cold, drizzly Tuesday, the ice-cream freezer hums like a lonely appliance in an empty room while umbrellas sell so fast the vendor can't make change quickly enough. Neither business is "better." They're just out of phase with each other.

That simple fact—*out of phase*—is the whole trick of diversification.

The point isn't to own a bunch of things. The point is to own things that take turns being disappointing.

Most investing mistakes begin as a perfectly reasonable human impulse: "Why not just buy the best thing?" If you knew, with certainty, which asset would win the next 30 years, concentration would be logical. One basket, one egg, no nonsense. The problem is that investing is not a talent show where the winner is announced in advance. It's a weather system. And in a weather system, the most dangerous belief is "this is the season forever."

Diversification is what you do when you accept a hard truth: you can be smart and still be surprised.

### Why ``more bets" isn't the same as diversification

Owning 50 stocks can still be one bet. Owning three different tech ETFs can still be one bet. Owning the same economic exposure in five different wrappers is like buying umbrellas from five different stores and calling it "multiple income streams."

Diversification is about *independence of outcomes*. You want pieces of your portfolio to respond differently to the same world. Ideally, when one piece is sulking in the corner, another is doing its job.

This is where the vocabulary matters, because finance uses a very specific word for "do these things tend to move together?" The word is *correlation*. Correlation runs from $-1$ to $+1$:

- $+1$: two assets move together, like synchronized swimmers.

- 0: no consistent relationship; sometimes together, sometimes not.
- $-1$: they move opposite each other, like a seesaw.

You don't need to memorize the scale. You just need to remember the feeling: high correlation means your portfolio has fewer true shock absorbers than it appears.

## The ``free lunch" claim (and the fine print)

Diversification gets called the only free lunch in finance because, in the math, you can sometimes lower risk without lowering expected return. That sounds like a violation of the laws of investing, where everything good is supposed to cost something.

The cost is not in the spreadsheet. The cost is in your ego.

Diversification means you will always own something embarrassing. Always. At any given moment, one of your holdings will look like the dumbest thing you've ever bought. If every part of your portfolio is performing well simultaneously, you don't have diversification; you have a trend.

The free lunch is real *in expectation*—when you don't know the future—but it is emotionally expensive because it forces you to tolerate being half-wrong all the time. That's the admission price. The meal is still free.

## A tiny bit of math that changes the whole picture

Here's the surprising claim: adding a *volatile* asset can make your *portfolio* less volatile.

That sentence sounds like a riddle until you see what portfolio risk actually depends on. It's not just "how wild is each asset?" It's also "do they get wild at the same time?"

If two assets move differently enough, their bumps partially cancel. The portfolio's ride smooths out even if the individual vehicles are bumpy. Think of it like noise-canceling headphones: they don't remove sound by making the world quieter; they remove sound by adding a carefully chosen opposite wave.

A portfolio is a noise-canceling machine for uncertainty. Correlation is the instruction manual.

If you like formulas, there's a compact way to say it for a two-asset portfolio. Let $w$ be the weight of Asset A, $1 - w$ the weight of Asset B, $\sigma_A$ and $\sigma_B$ their volatilities, and $\rho$ their correlation. Portfolio volatility is:

$$\sigma_P = \sqrt{w^2\sigma_A^2 + (1-w)^2\sigma_B^2 + 2w(1-w)\sigma_A\sigma_B\rho}.$$

The magic is in the last term: $2w(1 - w)\sigma_A\sigma_B\rho$. When $\rho$ is low, that term shrinks. When $\rho$ is negative, it subtracts. The portfolio stops acting like "a weighted average of two risks" and starts acting like "a machine that makes risks interfere with each other."

You don't have to compute this to benefit from it. You just have to stop thinking of diversification as "own many things" and start thinking of it as "combine different behaviors."

### Why smoothing the ride can increase compound returns

Even if diversification did nothing but make you feel better, it would be worth something. But it often does more than that: it can improve the *compounding* result.

Compounding isn't driven by average returns alone. It's driven by the path—especially the bad parts of the path. A portfolio that avoids deep holes spends less time climbing back to flat ground. A smoother return stream suffers less from the arithmetic penalty of big drawdowns.

There's another subtle piece that investors miss: volatility doesn't just annoy you; it can mathematically drag down geometric (compounded) returns. If returns bounce violently, the compounded growth rate tends to be lower than the simple average. Diversification, by reducing those violent bounces, can raise the growth rate of the whole machine even if the expected arithmetic return of each component doesn't change.

This is why "reducing risk" is not automatically the same as "reducing return." Sometimes reducing risk *protects* return, because it protects the compounding process that turns return into wealth.

### The practical meaning of correlation: what you want when you need it most

Correlation has a nasty habit: it behaves best in calm times and worst in crises.

When everything is fine, lots of assets look charmingly uncorrelated. When markets panic, correlations often rise. Investors sell whatever they can sell, not whatever they should sell. Liquidity becomes the common factor. In those moments, your portfolio can feel like it was diversified on Tuesday and concentrated by Friday.

This doesn't mean diversification is useless. It means you should be realistic about what it can and cannot do.

Diversification is not a force field. It is shock absorption. It doesn't promise you won't get hit. It tries to ensure that one impact doesn't total the car.

That realism also protects you from a common mistake: owning a portfolio that is "diversified" in a way that only helps during normal times. If your diversification plan collapses the moment you care about it, you don't have a plan; you have a fair-weather friendship.

### Diversification across what, exactly?

Most people diversify within a single category because it's familiar. They diversify "stocks" by owning many stocks. That helps, but it solves a narrower problem: the risk of any one company being a disaster.

True diversification often requires stepping across boundaries that feel psychologically uncomfortable:

- Different *asset classes* (stocks, high-quality bonds, cash-like instruments, real assets).
- Different *geographies* (your home country is not the whole world).
- Different *economic sensitivities* (growth surprises vs inflation surprises vs deflation scares).
- Different *styles/factors* (cheap vs expensive, stable vs flashy, etc.), even when they look silly for long stretches.

The goal isn't to collect stamps. The goal is to reduce the chance that one macro regime turns your entire portfolio into the same trade wearing different clothes.

A useful mental model is to ask: *what headline hurts this holding?* If the answer is the same headline for most of your portfolio, correlation is probably higher than you think.

### The behavioral advantage: fewer ``all-or-nothing" moments

A concentrated portfolio creates binary emotional states: genius or idiot, safe or doomed, euphoric or nauseous. Diversified portfolios are less cinematic. They generate fewer moments where you feel forced to make a dramatic decision.

That sounds boring. Boring is underrated.

The investor who compounds for decades isn't usually the one with the most exciting chart. It's the one who avoided the big self-inflicted wounds: panic selling, leverage blowups, whiplash strategy changes, and the corrosive habit of abandoning anything that feels temporarily humiliating.

Diversification reduces the frequency and intensity of humiliation. That is not a minor feature. It's the whole product.

### The opportunity cost: the part diversification advocates whisper about

There is a legitimate objection to diversification: it dilutes winners.

If U.S. large-cap growth stocks dominate for a decade, a globally diversified portfolio will look like it brought a salad to a pizza party. If one asset class has a historic run, diversified investors will spend years explaining why they "still believe in the process" while their neighbor points at a single ETF and smiles.

That's the psychological bill. You pay it in the currency of relative performance—*tracking error*, the pain of looking wrong compared to a popular benchmark.

The trick is to frame the choice correctly. Diversification is not trying to maximize bragging rights in the best decade. It's trying to maximize the probability that you stay invested through the worst decade.

If your plan requires you to correctly identify the future champion and never lose faith while holding it through brutal drawdowns, your plan is not an investing plan. It's a personality test.

### A concrete, unglamorous example: the ``shock absorber" role of high-quality bonds

Nothing tests the maturity of an investor like owning an asset that exists mostly to be boring.

High-quality bonds (think: government bonds, investment-grade bonds) often play this role. In many equity selloffs, they have historically tended to hold up better than stocks, sometimes even rising as investors seek safety. That low or negative correlation can make them a stabilizer.

But bonds also have their own bad years—especially when inflation jumps or interest rates rise. In those times, the stabilizer feels like it betrayed you. Investors then discover whether they wanted diversification or they wanted everything to go up.

A diversified portfolio is not a promise that nothing will hurt. It's an acceptance that different things hurt at different times, so you don't get wounded everywhere at once.

### How to tell if you're diversified without doing a PhD in statistics

You can do a quick audit with three questions that require more honesty than math:

1. *If inflation stays high for years, what happens to my portfolio?*
2. *If we get a deep recession, what happens to my portfolio?*
3. *If interest rates spike or stay elevated, what happens to my portfolio?*

If the honest answer to all three is "it probably gets crushed," you don't have diversification. You have a single macro bet. If the answers are meaningfully different—some pain here, some resilience there—you're building something that can endure different weather.

Another practical sign: in a diversified portfolio, you should almost always have at least one holding you're tempted to "clean up." The urge to simplify is often the urge to concentrate. That urge tends to spike right before diversification is most valuable.

### Actionable takeaway

Pick one existing holding in your portfolio that dominates your results—the position or asset class that would make you say "my portfolio lives or dies on this." Then add (or increase) a counterweight whose failure mode is different. Don't choose the counterweight because it has performed well recently; choose it because it tends to disappoint at *different* times. If you can explain, in plain language, the different headlines that would hurt each piece, you're doing correlation thinking instead of collection thinking.

## 3.3 Strategic Asset Allocation

If you've ever watched someone "get serious" about fitness, you know the moment. They buy shoes that look fast. They download an app with a confident name. They start debating supplements. All of that energy is real, and most of it is aimed at the wrong lever.

Investing has the same personality flaw. People obsess over the clever move—the perfect stock, the perfect entry point, the perfect macro call—when the lever that does the heavy lifting is boring enough to be mistaken for bureaucracy. Strategic asset allocation is that lever. It's not a trade. It's not a prediction. It's the default setting you return to when your brain is loud.

Call it your *Policy Portfolio*: the mix of assets you would choose if you were calm, sober, and immune to headlines. The purpose of a Policy Portfolio isn't to win every year. It's to keep you solvent, invested, and emotionally intact long enough for compounding to do its slow violence in your favor.

### The portfolio you actually own is your behavior

A portfolio is often described as a set of holdings. In practice, it's a set of *promises* you make to your future self.

You're promising that you'll keep contributing during boring years. You're promising you won't liquidate after a gut-punch drawdown. You're promising you won't "temporarily" go to cash and then forget to come back. Those promises are not kept by intelligence. They're kept by architecture.

Strategic asset allocation is architecture because it decides how often you'll be tested, and how hard. The earlier idea—volatility as the cover charge, drawdown as the cliff—now gets a practical implication: your Policy Portfolio is the cliff-height you are agreeing to live with.

It is also the cleanest way to avoid the most common investor tragedy: building a portfolio that *requires* emotional heroism. A good Policy Portfolio makes heroism unnecessary most of the time.

## ``Over 90%" and the story people tell themselves

You'll often hear a striking claim: asset allocation explains over 90% of return variability. It's usually delivered like a mic drop, the way someone might say "diet is 90% of weight loss" right before posting a picture of grilled chicken.

Here's the useful way to understand that claim without turning it into a slogan.

The day-to-day and year-to-year behavior of a portfolio is dominated by its big exposures: how much stock-market risk you're taking, how much interest-rate risk you're taking, how much inflation sensitivity you're taking, how much credit risk you're taking. Those exposures are mostly set by the mix of asset classes. Once you choose "mostly stocks" or "half stocks, half bonds" or "stocks plus real assets," you've already chosen the shape of most of your future experience.

Security selection can matter. Manager skill can matter. Costs and taxes can matter a lot. But if your policy mix is 90% gasoline and 10% brakes, it doesn't matter which brand of steering wheel you bought. The car has a personality now.

A crucial nuance: this doesn't mean asset allocation explains 90% of your lifetime wealth outcome. It means it explains a huge share of the *ups and downs you will live through*. And those ups and downs are what shape your behavior. Behavior, in turn, shapes lifetime wealth.

So even if you don't care about academic attribution studies, you should care about the implication: your Policy Portfolio largely decides which mistakes you'll be tempted to make.

### The core question: what is this money *for*?

Strategic allocation starts with an unglamorous question that most investors skip because it feels like paperwork: *what job is this money supposed to do?*

A Policy Portfolio for money you need in three years should feel almost insultingly conservative, because its job is not to be impressive; it's to be there. A Policy Portfolio for money you won't touch for 30 years can afford to be more volatile because time is the shock absorber.

The market will constantly try to seduce you into acting as if all dollars are the same. They're not. Dollars have timelines. Timelines have required stability. And required stability should dictate allocation more than your opinions about the Federal Reserve.

A practical way to frame it is to separate money into buckets by purpose:

- *Soon money*: near-term spending, known obligations, emergency reserves.
- *Later money*: retirement, long-horizon goals, legacy.
- *Never money*: money you truly might not spend (or you want to treat as an endowment for your future self).

These buckets can live in one account or many; the accounts are logistics. The allocation is philosophy.

### Risk tolerance is not a personality trait; it's a system constraint

Most advice treats risk tolerance as if it were a stable part of your identity, like being an introvert. That's not how it behaves.

Risk tolerance is conditional. It depends on:

- How secure your income is,
- How many obligations you carry,
- How much cash you have,
- How recently you've been hurt,
- Whether your friends are bragging.

In other words, risk tolerance is closer to blood sugar than to personality. It fluctuates.

That's why a Policy Portfolio should be set like a speed limit, not like a mood. You don't set the speed limit based on how excited you feel in the morning. You set it based on what happens when things go wrong.

The cleanest definition is behavioral: your risk tolerance is the maximum drawdown that you can survive *without changing your plan.* Not without feeling bad. Feeling bad is included. The question is whether you'll still execute.

### Choosing the big levers: the policy mix

The Policy Portfolio is usually expressed as percentages. It sounds simple: 60/40, 80/20, 100/0, risk parity, endowment-style. Those labels are useful shorthand, but the real work is understanding what you're buying with each choice.

### Equities: the engine (and the source of most nausea)

Stocks are ownership claims. Over long horizons, they've tended to be the portfolio engine because you're participating in business profits and growth. The cost is that stocks are not obligated to behave politely. They can fall far, stay down for a while, and make smart people doubt their sanity.

A higher equity allocation is not just "more return." It's a higher probability of being tested by deep drawdowns. That test is not theoretical. It's a calendar event, just not one you can schedule.

### High-quality bonds: the ballast (until they aren't)

High-quality bonds and cash-like instruments often serve as ballast: they dampen the portfolio's movement and can provide dry powder for rebalancing after equity declines. They also make your portfolio less dependent on selling stocks at bad times to fund life.

But ballast is not magic. Bonds can have ugly stretches, especially when inflation surprises or rates rise. The correct response is not to declare bonds "dead." The correct response is to recognize what role you hired them for. If you hired them to reduce equity crash risk and provide liquidity, judge them on that job, not on whether they look heroic next to a roaring stock market.

### Real assets and diversifiers: the specialists

Assets like inflation-linked bonds, commodities, and real estate can act as diversifiers—sometimes. They are specialists, not universal solutions. They may help in inflationary regimes or provide different sensitivity than stocks and nominal bonds. They can also be volatile, politically frustrating, and expensive to hold in the wrong wrapper.

Strategic allocation to these assets is less about chasing "alternative return" and more about acknowledging that the world has multiple kinds of bad years. You're designing for more than one villain.

### Three archetypes (not as rules, as mirrors)

It helps to see common Policy Portfolios as archetypes rather than prescriptions. The goal is to find the version that matches your constraints, not to join a tribe.

*The Classic Balanced Portfolio (e.g., 60/40).* Often used as a middle-of-the-road starting point. Its appeal is not that it's perfect, but that it has a long history of being *survivable.* It tends to reduce drawdowns versus all-equity portfolios while still allowing meaningful growth.

*The Aggressive Growth Portfolio (e.g., 80/20 or 100% equities).* This is honest about what drives wealth for long horizons: equity exposure. It can work beautifully for investors with stable incomes, long timelines, and the temperament to endure serious drawdowns without flinching. The danger is not the drawdown itself; it's discovering in real time that your temperament was imagined.

*The "Endowment-ish" Portfolio (diversified across multiple risk sources).* Often includes meaningful exposure beyond just stocks and nominal bonds. The hope is to diversify across economic regimes. The danger is complexity: higher fees, harder rebalancing, and the constant temptation to abandon the least-loved sleeve at exactly the wrong time.

The right archetype is the one you can hold through humiliation.

### The hidden decision inside every allocation: concentration vs. diversification of *risks*

People think strategic allocation is about choosing asset *classes*. More deeply, it's about choosing *risk sources*.

A simple stock-heavy portfolio is concentrated in "growth risk" and "liquidity panic risk." A stock-plus-bond portfolio spreads the risk across equity risk and interest-rate risk. A portfolio with additional diversifiers may spread across inflation sensitivity and other economic exposures.

The earlier lesson about correlation belongs here: you're not selecting ingredients for taste; you're selecting ingredients for how they behave when the kitchen catches fire.

So when you set your Policy Portfolio, try describing it without using asset-class names at all. Describe it in sentences like:

> "I'm comfortable having most of my long-run outcome tied to corporate profits, but I want enough high-quality ballast to avoid being forced to sell after a crash."

or

> "I worry about inflation regimes enough that I want some explicit inflation sensitivity, even if it looks silly in calm years."

If you can't describe your allocation in plain language, you're probably not allocating; you're collecting.

### The most underrated part: rebalancing as the enforcement mechanism

A Policy Portfolio without rebalancing is a policy without enforcement.

Markets drift your weights. If stocks soar, you silently become more aggressive. If stocks crash, you silently become more conservative. Left alone, your portfolio tends to buy high and sell low through inertia, which is a remarkably human way to fail at a mathematical game.

Rebalancing is the act of restoring your intended risk. It's the most mechanical way to behave rationally without requiring constant rational thought.

There are two common approaches:

- *Calendar rebalancing*: check and rebalance on a schedule (e.g., annually).
- *Threshold rebalancing*: rebalance when an asset drifts too far from its target (e.g., stocks deviate by more than 5 percentage points).

Neither is perfect. The point is to have a rule that prevents "I'll rebalance when it feels safe," because it never feels safe when rebalancing matters.

Rebalancing also ties back to drawdown control. In a diversified portfolio, rebalancing is how you harvest the fact that different pieces misbehave at different times. Without rebalancing, you're letting correlation do whatever it wants while you watch.

### Why ``set it and forget it'' is both right and wrong

Strategic allocation is not a market-timing call. You don't change your Policy Portfolio because a pundit sounded persuasive. You don't change it because the last 12 months were exciting. You don't change it because your neighbor had a great year.

But "set it and forget it" can become a lazy mantra if it prevents legitimate updates. Life changes are real. If your time horizon shrinks, if your liabilities grow, if your income becomes less stable, if you approach a spending phase, the portfolio's job changes. A new job may require new tools.

The right mental model is: *stable policy, responsive to life*. Not responsive to news.

One of the most useful habits is an annual "life audit" rather than a "market audit." Ask: has my timeline changed? Have my obligations changed? Has my need for liquidity changed? If the answer is no, the policy stays. If yes, the policy can evolve without it being a reaction.

### A small table that forces clarity

Sometimes the fastest way to make the decision real is to write down what each asset sleeve is *for*. Not what it is. What it's for.

| *Sleeve* | *Job description* | *When it will feel awful* |
|---|---|---|
| Equities | Long-run growth engine; participation in profits and productivity | Recessions, crashes, long bear markets; when you feel "behind" |
| High-quality bonds / cash-like | Stability, liquidity, spending buffer; potential ballast in risk-off periods | Inflation spikes; rate increases; when stocks are booming and bonds look pointless |
| Diversifiers (optional) | Reduce dependence on a single economic regime; hedge specific risks (e.g., inflation) | Long stretches of underperformance; when the hedge looks unnecessary |

That last column is the secret. A Policy Portfolio isn't defined by what looks good on a backtest. It's defined by what you can hold when its job requires it to be unpopular.

### Actionable takeaway

Write your Policy Portfolio in one sentence and one set of percentages. The sentence must describe the *job* (growth, stability, liquidity, regime resilience) and the *behavioral constraint* (the drawdown that would make you break). Then set a rebalancing rule you can follow without discretion (calendar-based or threshold-based) and put it on your calendar today. If your allocation isn't written down and enforced, it isn't policy—it's vibes.

## 3.4 Optimization and the Efficient Frontier

If you want to watch a room full of rational adults become romantics, put a portfolio optimizer on the projector.

Someone types in a list of funds, clicks a button, and out pops a number with the aura of destiny: *the optimal allocation*. The room leans in the way people lean in at a fortune teller. There's relief in it. A world of uncertainty just got reduced to a clean solution with two decimal places. The future has been tamed by math.

And then, quietly, the future ruins the math.

Optimization is one of the most beautiful ideas in finance, and one of the easiest ways to accidentally build a portfolio you can't live with. It offers a seductive promise: give me expected returns, volatilities, and correlations, and I'll find the portfolio that squeezes the most return out of each unit of risk. The problem is that those inputs are not facts. They're estimates. And estimates are where confidence goes to die.

The efficient frontier is real. The perfect portfolio is a mirage. The craft is learning the difference.

### The efficient frontier: the menu of ``best possible" trade-offs

Start with the simplest non-trivial idea: you can't maximize everything at once.

If you want higher expected return, you usually accept more risk. If you want less risk, you usually accept lower expected return. The efficient frontier is the boundary of those trade-offs. It's the curve that says: *for each level of risk, here is the highest expected return you can reasonably target, given the ingredients you allow yourself to use.*

Portfolios that sit *below* that curve are inefficient: they take too much risk for the return they expect, or earn too little return for the risk they take. Inefficiency sounds like a moral failing, but it's usually just an accident of composition. You picked assets that overlap too much, or you left out something that diversifies, or you are holding extra cash without realizing it.

If you have ever wondered why two investors can own "diversified portfolios" and yet one sleeps better *and* ends up richer, the efficient frontier is often the explanation. One portfolio sits closer to the frontier. The other wastes risk.

This idea was revolutionary when it was formalized in the early days of modern portfolio theory because it reframed investing from "pick good stocks" to "design good combinations." It said the portfolio is the product. Not the holdings.

## Risk and return are not opposites; they're coordinates

The efficient frontier lives on a two-axis map: risk on the horizontal axis, expected return on the vertical axis. That framing is powerful because it forces honesty.

You cannot argue with a chart that says: "This portfolio has the same expected return as that one, but with more expected volatility." Why would you choose the bumpier ride if you aren't being paid for it?

But the chart also hides a dangerous assumption: that risk can be summarized by a single number, typically volatility. That's a convenient proxy for the day-to-day wiggle, but investors don't fail because they experienced wiggle. They fail because they experienced the kind of sustained loss that makes them abandon the plan. The frontier is a map, not the territory.

Still, as maps go, it's a good one. It reminds you that portfolio design is an engineering problem: the goal is not to find a magic asset; it's to find a combination that doesn't waste risk.

### Where the optimizer shines: removing obvious stupidity

Before criticizing optimization, it's worth giving it its due.

A good optimizer can do a few things extraordinarily well:

- It can reveal when your "diversified" portfolio is actually a pile of highly correlated bets.
- It can show you the cost of constraints ("I refuse to own bonds" is a constraint with a price tag).
- It can make trade-offs explicit, rather than leaving them as vibes.

It's particularly useful as a diagnostic tool. Like a medical scan, it can show you structure you can't easily see by eyeballing holdings.

The danger begins when you treat the scan as a prophecy.

### The optimizer's fatal flaw: it believes your inputs

An optimizer is not smart. It is obedient.

It takes your expected returns and treats them as if they were gravity. It takes your correlations and treats them as if they were laws of physics. It takes your volatility estimates and treats them as if they were stable personality traits.

But expected returns are not known. Correlations are not stable. Volatility clusters. Regimes change. Entire economic eras flip the sign on relationships investors thought were permanent.

In other words: the optimizer is a high-powered machine that amplifies whatever beliefs you feed it. If you feed it noise, it will manufacture confidence out of noise.

This is why optimizers often produce allocations that look like they were designed by a mischievous intern:

- 63% in an obscure asset with a tiny historical edge,

- 0% in a popular asset because its recent history looks slightly worse,
- extreme concentration in whatever happened to look best in the sample period.

The optimizer is not telling you what will work. It is telling you what would have worked *if the past were a trustworthy forecast and your estimates were perfect*. That "if" is the whole story.

## The rearview mirror problem: precision about the wrong thing

Historical data is seductive because it's abundant and concrete. You can download decades of returns and correlations in minutes and feel like you're holding reality.

But finance has a cruel joke built in: the more precisely you estimate a number from history, the more you're tempted to believe it will hold. Markets, being markets, specialize in breaking what you come to rely on.

A classic example is correlation. In calm markets, stocks and bonds may show a pleasing diversifying relationship. In inflationary regimes, that relationship can weaken or flip. The optimizer, trained on yesterday's climate, is blind to tomorrow's weather.

Expected returns are even worse. Volatility can be estimated with some statistical stability. Expected excess returns are notoriously hard to estimate precisely. Small differences in assumed expected returns can lead to huge changes in "optimal" weights. That's not a bug. It's the math doing what it was asked to do: if you tell it Asset A has a slightly better expected return, it will try to stuff the portfolio with Asset A to harvest that edge.

This sensitivity is why optimization often feels like it produces fragile portfolios. A tiny change in inputs yields a dramatic change in outputs. Fragility is not a theoretical problem. Fragility is how you end up abandoning a portfolio the moment it underperforms, because it never had a stable narrative to begin with.

### Optimization error is not symmetric: it punishes confidence

When you make an estimation mistake, the consequences are not evenly distributed.

If you accidentally underestimate the volatility of an asset, you may allocate too much to it and discover the truth at the worst possible time. If you overestimate diversification benefits, you build a portfolio that looks stable on paper but behaves like a single bet in crisis. If you overestimate expected return, you take on risk without being paid.

The optimizer tends to push you toward the edges—the portfolios that look mathematically best given the assumptions. When the assumptions are wrong, the edges are where you fall off the cliff.

This is the emotional mismatch: optimization tempts you into maximum confidence (a precise allocation), while long-duration investing requires maximum humility (a portfolio that survives being wrong).

### The ``perfect" portfolio is often uninhabitable

Even if you could identify the true efficient frontier in advance, there is a separate question no optimizer can answer: *can you hold it?*

A portfolio is not optimal if it causes you to change strategies at the wrong time. A portfolio is not optimal if it requires you to ignore all social comparison, all career risk, all headlines, all regret. A portfolio is not optimal if it has a theoretically superior Sharpe ratio but a practically unbearable tracking error.

This is where the map meets the human.

An optimizer might love a portfolio that has a weird combination of assets that historically had low correlation and decent returns. You might hate it, because it looks nothing like what "a normal investor" owns. You'll be asked to defend it to yourself every year it lags the popular benchmark. That defense is work. Work is a cost. Costs reduce returns because they increase the probability of abandonment.

The compounding machine doesn't care whether your portfolio was optimal on a chart. It cares whether it stayed intact.

### How professionals make optimization usable: constraints and humility

In practice, optimization is rarely used in its pure, unconstrained form for exactly these reasons. Professionals turn the optimizer from a dictator into an assistant.

Common ways they do that:

- *Constraints*: limit how large any one allocation can be; require diversification across categories; prevent extreme leverage or concentration.
- *Robust inputs*: shrink extreme estimates toward more conservative values; use long-horizon assumptions rather than short-window history; blend multiple sources of expected return estimates.
- *Priors and judgment*: start with a sensible Policy Portfolio and let optimization suggest small tilts rather than wholesale redesign.

These techniques all share a philosophy: the optimizer is not allowed to express high confidence, because high confidence is not warranted.

This is not an admission of defeat. It's an admission that markets are not lab experiments.

### A better way to use the frontier: find the neighborhood, not the address

The frontier is useful as a *region* rather than as a pinpoint.

Instead of asking "what is the optimal portfolio?", a more durable question is: *what is an efficient neighborhood of portfolios that are broadly reasonable, diversified, and robust to estimation error?*

Think of it like choosing a home in a city you don't fully know. You don't need the single best address. You need a good neighborhood with reliable utilities, safe streets, and a commute you can tolerate. The optimizer can tell you where the bad neighborhoods are (inefficient portfolios that waste risk). It cannot tell you which exact house you will love in ten years.

That shift—from perfection to robustness—is one of the biggest upgrades a serious investor can make.

### The efficient frontier is conditional: change the ingredients, change the frontier

One of the most overlooked facts about the efficient frontier is that it depends on what you allow in the kitchen.

If you only allow U.S. stocks, your frontier will look one way. Add high-quality bonds, and the frontier typically shifts. Add international equities, real assets, or alternative risk premia, and the shape can change again. Add a risk-free asset (cash or short-term government bills), and you create a new anchor point for combining risky and risk-free exposures.

This is a powerful insight because it reframes the role of diversification. You're not diversifying to collect more tickers. You're diversifying to expand the set of efficient possibilities.

But this also means you can fool yourself by overfitting the ingredient list. If you add enough exotic assets, you can often manufacture an impressive-looking historical frontier that's really just a monument to data-mining. The more ingredients you add, the more degrees of freedom the optimizer has to hallucinate a perfect past.

The frontier gets smoother. Reality gets messier.

### A practical thought experiment: the optimizer vs. the Policy Portfolio

Imagine two investors:

Investor One builds a Policy Portfolio based on goals, time horizon, and behavioral limits. It's diversified, rebalanced, and intentionally boring. It is not optimized to the third decimal place. It is designed to be held.

Investor Two runs an optimizer monthly. They chase the newest "optimal" weights as estimates update. They feel scientific. They also live inside constant regime-switching, turnover, and second-guessing. Their portfolio is always a little different from last month's self.

Who compounds better?

The second investor might win in a world where inputs are stable and transaction costs are zero and emotions don't exist. The first investor tends to win in the world we actually inhabit: one where the biggest edge is staying invested, keeping costs low, and avoiding the catastrophic decision.

Optimization can be a tool inside strategic allocation. It cannot replace strategic allocation.

### Actionable takeaway

Use an optimizer (or any efficient-frontier tool) only as a *stress test*, not as an oracle. Start with a written Policy Portfolio you can hold through embarrassment. Then ask the optimizer one question: "Am I obviously inefficient given my assumptions?" If the tool suggests extreme allocations or dramatic month-to-month changes, treat that as a warning about input uncertainty—and respond by adding constraints, simplifying assumptions, and prioritizing robustness over precision. The goal isn't the perfect point on the curve; it's a portfolio that stays on the map when the weather changes.

## 3.5 Factor Premia: Beyond the Averages

A strange thing happens when you start paying attention to how investors talk about stocks. They speak as if "the stock market" is a single organism with a single personality: it goes up, it goes down, it's risky, it's rewarding. That's convenient—and incomplete.

The market return is not one flavor. It's more like a stew. Different ingredients contribute different tastes, and sometimes one ingredient dominates for years. If you only think in terms of "stocks" and "bonds," you're treating the stew like a glass of water. It will hydrate you, sure, but you won't understand why it tastes the way it tastes.

Factor premia is the name for those ingredients: systematic drivers of returns that show up again and again across many securities. They're not secret trading tricks. They're more like recurring patterns in human behavior and market structure that can be packaged into rules.

The appeal is obvious. If a simple, repeatable tilt can improve long-run returns or improve risk-adjusted returns, why settle for the plain-vanilla market portfolio?

The cost is less obvious, and it's where investors usually fail: factor investing asks you to endure long stretches of looking wrong while still being, in a sense, right.

#### What a ``factor'' actually is (without the academic throat-clearing)

A *factor* is a characteristic that explains why groups of stocks (or bonds) tend to earn different returns over time. It's not a story about one company. It's a story about a *type* of company.

Think of it like this: the market is a giant voting machine that reprices securities every day. Factors are the recurring ways the voting machine seems biased. Not necessarily in a corrupt sense—in a human sense. Investors overpay for glamour. Investors underpay for boring resilience. Investors chase what has worked recently. Investors panic at uncertainty. Those patterns, when persistent enough, can create return differences you can try to harvest.

The base factor is simply *market beta*: owning stocks at all. Beta is the return you get for taking the market's big, obvious risk—the risk that the whole system can suffer together.

Factor premia begins when you say: "Okay. But within stocks, are there systematic differences I can tilt toward?"

That question leads to an entire universe of "smart beta" products, academic papers, and arguments at dinner parties.

## Why factor investing exists at all: two explanations that can both be true

It's tempting to want one clean reason factors work. You won't get it. You get two families of explanations, and the truth may be a blend.

*Behavioral explanation:* investors are predictably irrational. They overreact, underreact, chase stories, hate embarrassment, and prefer lotteries to slow wealth. Those behaviors can push prices away from fundamentals in systematic ways. If prices get pushed far enough, expected returns change. A factor premium is, in this view, a reward for being willing to consistently do what feels uncomfortable.

*Structural explanation:* markets have frictions. Some investors are constrained (can't use leverage, can't short, can't hold illiquid assets, must hug a benchmark, must avoid career risk). Those constraints create predictable supply-and-demand imbalances. A factor premium is, in this view, compensation for providing something the market can't easily manufacture under those constraints.

Both stories imply the same practical message: factor premia, if they exist, are not free money. They are payment for swallowing a specific kind of discomfort that most investors refuse to swallow.

## The three celebrity factors: Value, Momentum, Quality

There are many proposed factors. Some are real. Some are probably artifacts of data mining. Some are real but too crowded, too expensive, or too hard to implement well. For a working investor, it's more useful to understand a small set deeply than to collect factor names like baseball cards.

### Value: "cheap" stocks

*Value* means buying stocks that are cheap relative to some measure of fundamentals—earnings, book value, cash flows, sales. The underlying intuition is almost offensively simple: if you pay less for the same dollar of business, your future return should be better.

So why doesn't everyone do it?

Because cheap things are often cheap for a reason. Value stocks are frequently companies with problems: declining industries, ugly headlines, questionable management, old-economy vibes. They don't feel like the future. They feel like compromise.

The value premium, when it shows up, often feels like the reward for buying what you're not proud to own. That pride component is not a footnote. It's the mechanism. If value investing felt good all the time, it would be arbitraged away by everyone with a pulse and a brokerage account.

Value also teaches a brutal lesson about patience. It can underperform for years. Not months. Years. If you can't tolerate a long stretch of being the person who owns "the wrong stuff," you are not a value investor. You are a tourist who visited value for a weekend.

## Momentum: "winners keep winning"

*Momentum* means tilting toward stocks that have gone up recently and away from stocks that have gone down recently (usually defined over the past 6–12 months, with some implementation details). This offends the part of our brain that learned "buy low, sell high" in kindergarten.

And yet momentum has been one of the most persistent anomalies documented across markets and geographies. It's often explained as underreaction and herding: information gets incorporated into prices gradually, investors chase what's working, and trends persist longer than feels reasonable.

Momentum's psychological cost is different from value's. Value makes you feel old-fashioned. Momentum makes you feel like a trend-chaser. It can force you to buy things you think are overpriced and sell things you think are "due for a bounce." It can also crash hard when trends reverse violently. Momentum can make you money and make you feel foolish in the same year.

The premium, if you capture it, is partly payment for tolerating that whiplash without deciding you've been duped.

## Quality: "boring excellence"

*Quality* is less about cheapness and more about business strength: profitability, stable earnings, prudent balance sheets, good capital allocation, resilience. Quality is the factor version of preferring the sturdy bridge over the flashy one.

Quality tends to appeal to investors because it sounds like common sense. But it isn't automatically a free upgrade. If everyone decides they love "high quality," the price of quality rises and the expected return can fall. The lesson is uncomfortable: even excellence can be a bad investment if you overpay.

Quality's tracking-error pain often shows up in frothy markets. When speculative, unprofitable, story-driven companies surge, quality can look slow and out of touch. It's hard to be the person eating vegetables at a party where everyone else is doing shots.

## Smart beta: rules instead of stories

"Smart beta" is a marketing term, but it points to something important: factor investing is usually implemented with rules.

Instead of saying, "I believe Company X is undervalued," you say, "I will own a diversified basket of stocks with low valuation ratios," or "I will own stocks with strong recent performance," or "I will own stocks with high profitability and conservative balance sheets."

Rules have two advantages:

- They reduce the temptation to rationalize every decision.
- They make the exposure persistent, which matters because factor premia are not reliable on a human attention span.

Rules also have a disadvantage: they don't care about your feelings. When a factor looks terrible, the rule keeps doing the thing that looks terrible. That's the point—and that's why many investors abandon factors at the bottom of their relative performance.

## Factors as a language for portfolio architecture

Once you start seeing factors, you realize something liberating: asset classes are not the only meaningful buckets.

Two equity funds can both be "stocks" and yet behave very differently because one is loaded with value exposure and the other is loaded with growth exposure. Two bond funds can both be "bonds" and yet behave differently because one is taking lots of credit risk and the other is mostly interest-rate risk.

Factors give you a second coordinate system, like switching from street addresses to GPS. Asset classes tell you the neighborhood. Factors tell you the exact location.

That matters because it makes hidden concentration visible. Many investors think they are diversified because they own multiple funds, but those funds may share the same underlying factor bets. You can end up with a portfolio that is secretly one trade: "I am long expensive growth and low rates," or "I am long high beta and optimism."

Factor awareness turns that kind of accidental concentration into a conscious choice.

### The real cost: tracking error is a form of social pain

The most important risk in factor investing is not volatility. It's *relative underperformance*. The pain of watching your portfolio lag a plain market index while commentators announce that "value is dead" or "momentum is broken" or "quality is overpriced."

That pain has a name: *tracking error*. Tracking error is simply the degree to which your returns differ from a benchmark. It's not inherently bad. It's the whole reason you might earn a premium. But it is psychologically vicious because it turns investing into identity.

If you buy a total market index and it underperforms, you can shrug and say, "The market had a bad year." If you buy a factor tilt and it underperforms, you feel personally responsible. You didn't just lose money. You lost *status*. You look like you outsmarted yourself.

This is why factor premia, even if they exist, are hard to harvest. The market doesn't just charge you with volatility. It charges you with embarrassment.

### The implementation trap: factors are not one thing

"Value" is not a single definition. Neither is "quality." Momentum has many variants. Every index provider and fund sponsor has choices to make:

- Which metrics define the factor?
- How often do you rebalance?
- Do you use profitability screens?
- How do you handle sector biases?
- How do you control turnover and trading costs?

Two "value" funds can produce very different experiences. One might be deep value with lots of distressed names. Another might be value-with-quality, avoiding the worst balance sheets. The first can have bigger tracking error and deeper drawdowns. The second can be more palatable but may dilute the intended exposure.

This is where factor investing stops being a concept and becomes a craft. The premium, if any, is earned net of fees, trading costs, taxes, and the inevitable imperfections of real-world execution.

In taxable accounts, turnover-heavy strategies (momentum in particular) can be tax-hostile. Even if the pre-tax factor premium exists, the after-tax premium you personally experience may be much smaller. Factor investing is not just a return story; it's a plumbing story.

### A practical way to think about factor tilts: seasoning, not a new diet

Most serious long-horizon investors don't need to swing from "market index" to "factor maximalist." That's usually a recipe for regret.

A more robust approach is to treat factor tilts as seasoning. Your Policy Portfolio is still the meal. The seasoning can improve it, but it can also ruin it if you dump in the whole jar.

Seasoning has a practical advantage: it reduces the probability of abandonment. A modest tilt that underperforms is annoying. A full-factor portfolio that underperforms can feel like a referendum on your intelligence.

If you want to harvest factor premia, you must design around the likelihood that you will someday hate what you own.

### When factor investing is a good idea (and when it's self-sabotage)

Factor tilts tend to be more appropriate when:

- Your time horizon is long enough to survive multi-year droughts,
- You can keep costs and taxes under control,
- You can articulate, in advance, why you'll keep holding through underperformance,
- Your base portfolio is already sensible and diversified.

It tends to be self-sabotage when:

- You're using factors as a substitute for having an allocation policy,
- You need the money soon,
- You are chasing the factor that just had a great run,
- You can't tolerate being different from the benchmark.

The most common failure mode is not picking the wrong factor. It's buying the right factor at the wrong psychological moment: after a strong run, when it feels obvious and safe—right before the inevitable cold spell.

### Actionable takeaway

Choose *one* factor tilt you can defend in a bad decade—not in a good one. Write down a simple rule for yourself: the maximum tracking-error pain you're willing to endure (for example, "I will stick with this tilt through five years of underperformance"), and the maximum total allocation you'll devote to it (seasoning, not reinvention). Then implement the tilt in the cheapest, most tax-aware wrapper you have access to, and commit to evaluating it on a multi-year schedule rather than a monthly one. The premium, if it's real, is paid to investors who can keep holding when it stops being fun.

## 3.6 The Double-Edged Sword of Leverage

Leverage is the financial equivalent of a power tool. In the right hands, it builds faster. In the wrong hands, it removes fingers with the same efficiency.

The pitch is intoxicatingly simple: if you can earn 8% a year on your investments, then borrowing to invest more should earn you *more* than 8% a year. Multiply the exposure, multiply the result. It's the kind of reasoning that makes people feel clever because it is, in a narrow sense, correct.

The catch is that markets don't hand you returns in a straight line. They hand you a path. Leverage doesn't just multiply the destination—it multiplies every pothole along the way. And the potholes are where compounding is won or lost.

If diversification is the art of making your portfolio less fragile to surprise, leverage is the act of making your portfolio more sensitive to surprise. Sometimes that sensitivity is deliberate and disciplined. More often it's accidental: a margin account opened for convenience, an options position that quietly becomes the whole story, a leveraged ETF bought like it's a long-term holding because the ticker looks friendly.

The first job of leverage education is to replace the naive question—"what's the expected return?"—with the adult question: *what can go wrong before I get the chance to be right?*

## Leverage comes in costumes

Leverage isn't always labeled "borrow money." It shows up wearing different outfits, which is part of why people underestimate it.

*Margin* is the obvious costume: you borrow from your broker against your account, pay interest, and your positions are collateral. Your portfolio becomes a loan-backed machine.

*Leveraged ETFs* are a subtler costume: the fund uses derivatives and daily rebalancing to target a multiple of the *daily* return of an index ($2\times, 3\times, -1\times$, etc.). Many investors buy them as if they target a multiple of the *long-term* return. That misunderstanding is not a small detail. It's the whole trap.

*Options* are leverage with a clock attached. They are powerful precisely because a small premium can control a large notional exposure. They can also expire worthless with stunning indifference.

*Concentration* can be a form of implicit leverage. Owning one high-volatility stock with a story you love can create portfolio behavior that resembles leverage: enormous upside, catastrophic drawdown potential, and a strong chance of making a permanent decision in a temporary panic.

The mechanics differ, but the shared feature is this: leverage makes the path matter more than your thesis.

## Two distinct dangers: ruin and drag

Most people know the first danger: leverage can blow you up. If you borrow and your assets fall far enough, you can face forced selling, margin calls, or insolvency. That's the cliff.

But the more subtle danger is the one that quietly kills long-horizon plans: *volatility drag*, sometimes called *variance drain*. It is the slow leak that can cause a leveraged strategy to underperform even when the underlying investment goes nowhere.

This is what makes leverage so unintuitive. You can be directionally right over the long run and still lose because the path was hostile.

### The cliff: forced selling is not a theoretical risk

Leverage has a villain that unlevered investors rarely meet: the forced liquidation.

With margin, your broker is not your partner; it is your creditor. When the value of your collateral falls, the broker wants its loan to be safer. That can mean a margin call: deposit more cash or securities, or positions get sold. And they get sold when you least want them sold.

The market doesn't need to go to zero for this to happen. It only needs to fall enough, fast enough, to put your equity below maintenance requirements. And maintenance requirements can change. Brokers can tighten "house" rules. Volatility can rise. Liquidity can vanish. You can be right in the long run and still be removed from the game in the short run.

That is why leverage deserves the "risk of ruin" label more than almost any other tool. It adds a new failure mode: *timing forced upon you.* Compounding hates forced timing.

The psychological cost is also brutal. In a drawdown, an unlevered investor can choose to do nothing. A levered investor may be compelled to do something, often the worst something.

### The leak: volatility drag and why $2\times$ rarely means $2\times$

Now for the part that surprises people even after they've been investing for years: leverage can reduce long-term returns even if it increases exposure.

The intuition starts with a simple fact about compounding: losses hurt more than gains help. Not in a moral sense, in an arithmetic sense.

Suppose an asset goes down 10% one day and up 11.11% the next day. It's back to even. Fine.

Now imagine a daily $2\times$ leveraged version of that exposure. It goes down 20% and then up 22.22%. Do you get back to even?

No. You end up below where you started:

$$(1 - 0.20) \times (1 + 0.2222) = 0.8 \times 1.2222 \approx 0.9778.$$

You lost about 2.22% over two days even though the underlying ended flat.

That is volatility drag in its purest form: when returns are choppy, the multiplicative nature of compounding punishes the leveraged path. The more leverage you apply, the more the punishment grows.

This isn't a niche corner case. It's a structural feature of daily-reset leverage products and of leveraged portfolios that must maintain a constant leverage ratio by rebalancing as prices move. The mechanism is path dependence: the order of returns matters.

And because markets are often choppy—especially in stressed periods—volatility drag is not a rare event. It's a recurring tax.

### A sideways market that eats leveraged investors for lunch

The easiest way to see variance drain is not with a formula. It's with a simulation: a market that goes nowhere over time, but gets there in the most annoying way possible—up and down, over and over.

In that environment, an unlevered investor can finish roughly flat. The levered investor can finish meaningfully down, not because the market trended against them, but because the path was hostile to leverage.

That feels unfair until you remember what you bought: sensitivity to the path.

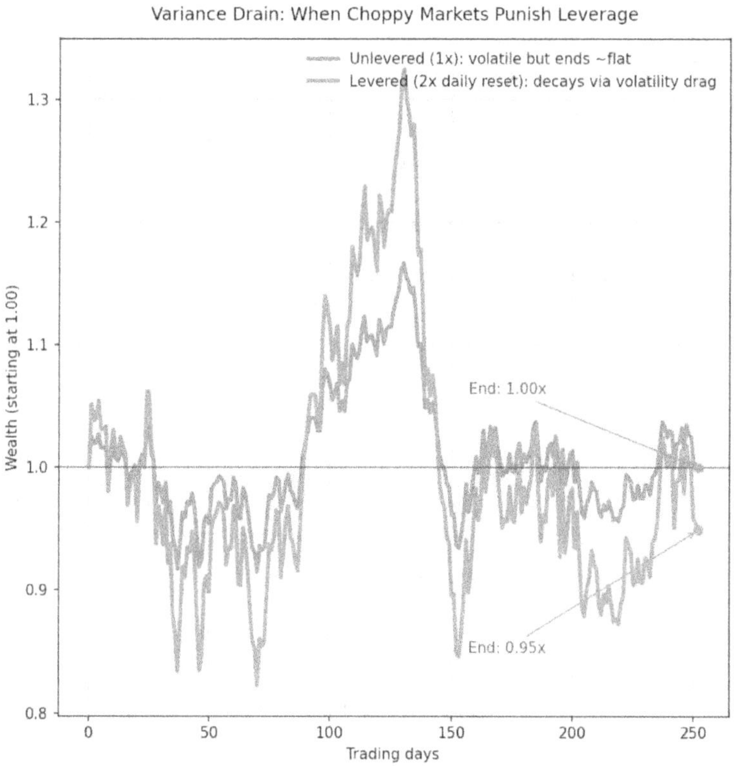


This is the kind of chart that makes investors argue because it attacks an intuitive belief: "If the index ends where it started, how could I lose?"

You lost because you didn't buy the index. You bought a daily-reset magnifying glass. And magnifying glasses don't just enlarge the picture; they concentrate heat.

### Path dependence: the sequence is the strategy

Leverage makes the order of returns more important. That's not philosophy; it's mechanics.

If you have leverage and you get hit early, your capital base shrinks, and every subsequent percentage gain is applied to a smaller number. Even if the market recovers, your levered position may not fully recover because the math of loss doesn't care about fairness.

This creates a nasty feedback loop:

- A decline reduces your equity.
- Reduced equity increases effective leverage (unless you rebalance).
- Higher effective leverage increases vulnerability to further declines.

In margin accounts, that loop can end with forced selling. In daily-reset products, that loop can end with decay that feels mysterious until you understand variance drain. In options, that loop can end with time decay taking over while you wait to be right.

Leverage turns "time in the market" into "time under constraint."

### So is leverage always bad? No. It's just rarely what people think they're buying.

There are disciplined uses of leverage that are closer to engineering than to gambling.

One rational motivation is to use modest leverage to scale a diversified, risk-balanced portfolio rather than concentrating into the riskiest assets. For example, an investor might prefer a portfolio with meaningful exposure to high-quality bonds and equities but use controlled leverage to reach a targeted overall risk level. In that framing, leverage is not a substitute for diversification; it's a complement to it.

Another rational use is tactical and time-limited: hedging, short-term positioning, or liquidity management by sophisticated investors who understand that they are managing a path-dependent object, not a simple bet.

But for most long-horizon investors, the most common reason leverage appears is more emotional: impatience. The desire to compress time. The hunger to turn a decade into a year. That hunger is understandable. It is also exactly what leverage punishes, because leverage makes you fragile to the very volatility you must live through on the way to long-run returns.

### A simple leverage checklist that prevents most disasters

You don't need to swear off leverage to respect it. You need rules that are strict enough to protect you from yourself.

Here are the questions that matter more than "what's the expected return?"

- *What is the maximum drawdown that would trigger forced selling?* If you can't compute it, assume it's smaller than you think.
- *Is this leverage daily-reset?* If yes, assume path dependence is a feature, not a bug.
- *Can I survive a 50% equity drawdown without selling anything?* If the answer is no, adding leverage is not *more aggressive*. It's structurally unsafe.
- *Do I understand the financing cost?* Borrowing isn't free, and financing rates can change when markets are stressed.

- *If this goes against me for six months, will I add, hold, or quit?* If the honest answer is "quit," leverage will likely turn that quitting impulse into a permanent loss.

The theme is always the same: leverage should be evaluated as a *survivability* question, not an *optimism* question.

### Actionable takeaway

If you are considering leverage, write down two numbers *before* you place the trade: (1) the maximum loss you will tolerate without adding money or changing your plan, and (2) the price level or portfolio value at which you would be forced to sell (by margin rules, product mechanics, or personal cash constraints). If those numbers are uncomfortably close, you don't have a leverage strategy—you have a liquidation strategy. The compounding machine can survive volatility; it cannot survive being kicked out of the market.

Chapter 4

# The Behavioral Firewall

*If investing were purely about math, librarians and accountants would be the richest people on earth. But the spreadsheet doesn't account for the feeling of your stomach dropping when the market dives 20%. This chapter builds a 'Behavioral Firewall'-a set of mental defenses designed to protect your compounding machine from its greatest threat: you.*

## 4.1 The Enemy Within: Behavioral Basics

Your portfolio can be perfectly diversified, low-cost, tax-aware, and engineered like a Swiss watch—and still get kneecapped by a 20-minute scroll through your brokerage app on a bad day. That's the uncomfortable premise: the greatest risk to long-duration wealth often isn't the market; it's the investor *inside* the investor.

The market is a modern environment. Your brain is not.

Evolution didn't design you to maximize *terminal net worth at age 75.* It designed you to survive the afternoon. If youŕe wired to avoid becoming lunch, you will be exquisitely sensitive to losses, threats, and social cues—and almost indifferent to slow, probabilistic rewards. That wiring was a miracle on the savanna. Itś a liability in a market where the main "predator" is volatility and the main "food source" is patience.

Picture two worlds.

In World A, danger is usually immediate and physical. When the bushes move, it might be wind, or it might be teeth. If you assume "wind" and youŕe wrong, you dont́ get a second chance. In World B, danger is often noisy but temporary. When your index fund drops 2% in a day, it feels like the bushes moved. Your body reacts as if teeth are nearby. But the correct action might be... nothing. Or even buying more.

That mismatch creates a permanent internal conflict: your biology wants certainty and safety *now*; compounding wants discomfort and discipline *for a long time*.

What makes this especially cruel is that your brain doesn't merely *think* about risk. It *experiences* risk. The same physiological machinery that spikes your heart rate in a near-miss car accident can light up when you see a streak of red numbers. Your body doesn't care that "this is just a screen." It cares that your status, security, and future feel threatened. The market's threat is abstract; your reaction is concrete.

And the market is very good at pressing those buttons.

Financial media is a threat-delivery system. It is engineered (not always maliciously, but inevitably) to convert randomness into narrative urgency: *Dow tumbles, stocks slammed, investors flee, bloodbath*. The words are not neutral. They're built to trigger attention, and attention evolved to prioritize danger. A headline that reads, "Expected long-run equity returns remain positive" does not survive in the wild. A headline that screams, "CRASH" does.

Your attention is valuable; your calm is fragile.

This is the first behavioral basic: you are not a spreadsheet with emotions as a software bug. You are an animal with a spreadsheet bolted on. The spreadsheet is useful, but it does not drive.

A second basic follows immediately: you don't experience market outcomes in dollars; you experience them as *changes relative to a reference point*. That reference point is usually a recent high-water mark: the day you bought, the day your account hit a milestone, the week your coworker started bragging about their "easy gains." Your brain doesnt ask, "Am I on track for my goals over 30 years?" It asks, "Am I up or down since *that moment that now feels like normal?*"

This is why a person can be up 120% over a decade and still feel miserable during a 12% drawdown. The loss is measured not against where they started, but against where they were last month when everything felt safe. The market, in other words, doesnt have to destroy your wealth to destroy your mood. It only has to move below your current reference point.

In the wild, reference points were adaptive. If youre used to finding water at a particular spot, and it dries up, you dont calmly average your long-term hydration. You panic and move. If youre used to safety in a familiar area, and predators appear, you dont "stay the course." You change course. Your brain is built to treat deviations from the recent normal as information-rich, urgent, and worthy of action.

Markets weaponize deviation.

Another ancient tool that becomes a modern trap is pattern-seeking. Humans are prediction machines. The world used to be full of repeating cause-and-effect relationships: clouds mean rain; tracks mean an animal passed by; a certain berry makes you sick. Finding patterns improved survival. Missing patterns killed you. So we evolved to see patterns even when they arent there—because false positives were cheap compared to false negatives.

In markets, false positives are expensive.

A chart goes up three days in a row and your mind writes a story: "momentum," "breakout," "the start of something." A chart goes down five days and your mind writes a different story: "something is wrong," "this time is different," "get out before it gets worse." Your brain is doing what it was built to do: convert a squiggle into a narrative.

But much of market movement is not narrative. Its noise, positioning, liquidity, shifting discount rates, or millions of independent decisions washing against each other like waves. The problem isn't that stories are always wrong. The problem is that stories feel *true* before they are proven true, and they come with a powerful action impulse attached.

The market tempts you with meaning. Your brain rewards you for finding it.

The action impulse is not a minor detail. It's the central danger. When humans feel threatened, action reduces anxiety. Doing something restores a sense of control. Selling "so you don't lose more" feels like turning the steering wheel away from an oncoming truck. Buying "before it's too late" feels like grabbing the last seat on a lifeboat. In both cases, action is emotionally soothing.

The market doesn't pay you for emotional soothing. It often charges you for it.

You can see this clearly in the simplest example: the difference between *risk* and *volatility*. Risk is the chance you permanently impair your ability to reach your goals. Volatility is the day-to-day and year-to-year wobble along the path. Your brain treats volatility as risk because in the ancestral world, volatility *was* risk. A sudden change in conditions often meant your survival odds just got worse.

In investing, volatility is frequently the price of admission.

Long-duration wealth is built by owning assets that are allowed to fluctuate while you keep your hands off the eject button. That is a bizarre behavior from an evolutionary point of view. It requires you to watch your "resources" shrink on paper and refrain from moving them into something that feels safer in the moment. It requires delayed gratification on a scale that nature rarely demanded: years, decades, an entire working life.

Nature trained you for the sprint. Markets reward the marathon.

There's another behavioral basic that stings because it's so personal: social proof. Humans are tribal. Being cast out used to be a death sentence; belonging was protection. So we evolved to track what other people are doing and to feel discomfort when we diverge. When everyone in your peer group is excited about a certain stock, sector, or strategy, not participating can feel like standing outside a warm house in winter.

FOMO is not just greed. It's social pain avoidance wearing a money costume.

This is why bubbles are rarely powered by dumb people. They're powered by normal social brains. The most dangerous stories aren't obviously ridiculous. They're plausible, emotionally resonant, and validated by other people's apparent success. The crowd becomes evidence. Price becomes proof. The rising chart becomes a kind of social permission slip: "It can't be wrong; look how many people believe it."

When you finally buy in, it often feels like relief. You're no longer the cautious outsider. You're back in the tribe.

Then, when the trade turns against the crowd, the emotional geometry flips. You don't just lose money; you lose *status*. You lose *self-image*. You lose the comfort of being aligned with the winners. Panic selling often isn't a calculation about discounted cash flows. It's a desperate attempt to stop feeling stupid and exposed.

Markets are one of the few places where feeling socially safe can be financially dangerous.

If all this makes you feel slightly doomed, there's good news hiding in plain sight. These reactions are common. They are not a personal moral failure. They are a predictable set of tendencies. In fact, the predictability is what makes them manageable—because anything predictable can be designed around.

But first, it helps to name the internal cast of characters.

One character is the *Alarm System*. It's fast, emotional, and binary. It thinks in categories: safe/danger, win/lose, now/never. It is allergic to ambiguity. It cannot be reasoned with in real time because its whole function is to act before reasoning finishes. When markets drop sharply, the Alarm System doesn't say, "Expected return may be higher now." It says, "This is how people lose everything."

Another character is the *Storyteller*. It hates randomness. It will happily glue together unrelated events into a coherent explanation because coherence feels like control. If the market falls the day after a scary headline, the Storyteller concludes the headline *caused* the fall and therefore more scary headlines imply more falling. The Storyteller rarely asks, "Was this already priced in?" It asks, "What is the plot?"

The third character is the *Accountant*. This is the rational part that understands diversification, long horizons, and probabilities. The Accountant is real, and it's the part of you reading a finance book for fun. But the Accountant is slow. It speaks softly. It often arrives late to the argument—and finds the Alarm System already halfway through a "protect us" trade.

A huge part of long-duration investing is not increasing IQ. It's giving the Accountant procedural power over the other two.

That brings us to another behavioral basic: most investing mistakes happen at *decision points*, not in spreadsheets. The numbers can be right and the actions can still be wrong. You can know that stocks have historically recovered from drawdowns and still sell in the middle of one. You can know that concentrated bets increase risk and still double down on your employer's stock because selling feels like betrayal. You can know that "time in the market beats timing the market" and still keep trying to time because *not trying* feels negligent.

The market often doesn't defeat your logic. It defeats your *identity*.

Identity is the silent driver. When you buy an investment, you rarely buy only an asset. You buy a self-concept. "I am a disciplined investor." "I'm the kind of person who saw this coming." "I'm prudent." "I'm bold." When prices move against you, the pain isn't just financial. It's a threat to that story.

This is why people cling to losing positions far longer than they intended. Selling is not simply admitting the thesis was wrong; it can feel like admitting *you* were wrong. And humans will pay surprising amounts of money to avoid that feeling. They will also pay surprising amounts of money to earn the opposite feeling: vindication. The desire to be proven right can turn a small position into a reckless one.

Evolution didn't just wire you to survive; it wired you to protect your place in the social hierarchy. Markets give that instinct infinite ways to express itself.

So what does "being rational" actually mean here?

It doesn't mean suppressing emotion. That's the most common misconception. Suppression works until it doesn't, and then it fails explosively. The goal is to anticipate the predictable emotional surges and build a structure where you don't have to win a white-knuckle internal debate every time the market moves.

Think of it like designing a building in an earthquake zone. You don't lecture the building about staying calm. You install shock absorbers. You use flexible materials. You design for the known forces. You respect physics. Behavioral finance is the same: you respect biology.

And you start by accepting one brutal truth: the market will always provide a reason to panic.

There will always be a recession risk, a war risk, an election risk, a valuation risk, an inflation risk, a tech risk, a "this is unprecedented" risk. Some of these risks will be real. Some will be overblown. The point is not to predict which is which in real time. The point is to recognize that your Alarm System treats *all* of them as "lion in the grass."

If you wait for a world with no reasons to worry, you will never invest. If you invest and then demand a world with no reasons to worry, you will never stay invested.

This is where the idea of an internal *firewall* becomes practical rather than poetic. A firewall isn't bravery. It's a boundary between a trigger and an action. It's the delay that prevents a reflex from becoming a trade. It's the rule that forces you to ask one more question before you hit "sell." It's the process that turns your emotional brain into a smoke alarm—useful for alerting you—rather than a pilot yanking the controls mid-flight.

Before the firewall can work, you need one more reframe that many investors never make: the market is not a morality play. It does not reward virtue on your timeline. It does not punish vice immediately. It does not care that you "did your homework." In the short run, outcomes are noisy and often unfair.

This is precisely why your brain struggles with it.

In the ancestral world, feedback loops were tight. If you touched fire, you learned quickly. If you ate the wrong thing, you learned quickly. Investing has delayed, ambiguous feedback. You can take a bad risk and get rewarded for years. You can make a prudent choice and look foolish for months. Your brain desperately wants to connect action to outcome. Markets refuse to cooperate.

So you substitute a different kind of feedback: emotional feedback. You judge decisions by how they feel. You judge success by the immediate result. And you become vulnerable to the most expensive habit in investing: changing a long-term plan in response to short-term sensation.

The antidote is not to become emotionless. It is to become *procedural*.

That means you deliberately separate two things most people blend together: *information* and *instructions*. Market moves are information. Headlines are information. Your fear is information. None of those things are automatically instructions.

Fear can be a signal—"this position is too big," "this plan is unclear," "this money is needed sooner than you thought." But fear is also often just the body doing body-things in response to volatility. The practical skill is learning to treat fear like a check-engine light: it warrants a diagnostic, not an immediate swerve across three lanes.

Hereś a concrete way to practice that diagnostic. The next time you feel urgency to make a trade, pause and write down (on paper, not in your head) the exact sentence completing this prompt:

> "I am about to take action because I believe that if I do not, ____________________________ will happen."

That blank is where your brain reveals its real model of the world. "Íll lose everything." "Íll miss the run-up." "Íll look stupid." "Íll never recover." "Íll regret it forever." Once itś written, it becomes discussable. You can test it. You can compare it to base rates (how often markets actually behave that way). You can ask whether itś a forecast or a feeling.

Writing it down turns a reflex into a hypothesis.

A second practice is to deliberately slow the stimulus-response loop. The market is fast. Your phone makes it faster. You doní need to be. If the trade is truly necessary, it will still be necessary tomorrow. If the trade is emotional, delaying it often makes it evaporate.

One of the most underrated skills in investing is the ability to sit with the uncomfortable sensation of "I should do something" and choose "I will do nothing" until you have a calmer mind. That isní passivity. Itś control.

Compounding rewards control.

*Actionable takeaway:* Create a personal "panic protocol" before the next drawdown arrives. Write three lines on a note you can actually access:

- A *cooling-off rule*: "No sells during market hours when I feel urgency. Re-check tomorrow."
- A *diagnostic prompt*: "Am I responding to volatility (noise) or to a permanent change in my plan?"

- A *default action*: "If I dont́ have a written reason tied to my long-term goals, I do nothing—or I add a small, pre-planned amount."

Keep it simple enough to follow when youŕe not thinking clearly—because thatś exactly when youĺl need it.

## 4.2 Loss Aversion and the Fear of Regret

A 1,000*gainisapleasantemailyouforwardtonoone.A*1,000 loss is a small personal scandal. It lingers. It changes your posture. It makes you check your account again, as if the numbers might apologize and reverse themselves while you watch.

That imbalance is not a quirk; it's architecture. In the psychology of money, losses don't simply "count" more. They *feel* more real. And because investing is one long sequence of feelings punctuated by occasional decisions, that asymmetry shapes behavior far more than any spreadsheet.

Prospect Theory—developed by Daniel Kahneman and Amos Tversky—starts with a rude observation about how humans actually make choices under uncertainty: we don't evaluate outcomes based on final wealth the way a textbook "rational agent" would. We evaluate changes relative to a reference point, and we treat losses as a different species of experience than gains. Same dollars, different universe.

The reference point is usually whatever makes you most emotionally vulnerable at the moment. Sometimes it's your purchase price. Sometimes it's the account peak from three weeks ago. Sometimes it's your friend's braggy screenshot. The reference point is not a financial concept. It's a psychological home base. When you move above it, you feel competent. When you move below it, you feel threatened—even if, in any objective sense, you're still doing fine.

This helps explain why markets can be so exhausting. The market doesn't just go up and down; it repeatedly drags your reference point around like a magnet. In a rising year, your "normal" quietly ratchets higher. A pullback that would have seemed trivial in January feels unacceptable by November. You didn't become less rational. Your reference point moved.

Loss aversion is the headline feature: the value function—the mapping from dollars to felt experience—is steeper for losses than for gains. Many people summarize it as "losses hurt about twice as much as gains feel good." Don't get hung up on the exact ratio; the important part is the shape. Your brain does not run on symmetric math.

Right here is the central tragedy of long-duration investing: the market's best long-run engine (owning risky assets for decades) requires tolerating periodic losses, but your brain experiences those losses as disproportionately catastrophic. The discomfort is not a sign you're doing it wrong; it's a sign you're human. Unfortunately, the market will happily accept "human" as payment.

The curve also bends. Gains show diminishing sensitivity: the difference between being up 1,000*andup*2,000 feels meaningful, but the difference between being up 101,000*and*102,000 barely registers. Losses do the same thing in the other direction: the first drop below your reference point is sharp pain; deeper losses still hurt, but the pain doesn't increase linearly. This curvature matters because it nudges people toward a particular kind of self-sabotage: they become desperate to avoid crossing below the reference point, then oddly numb once they're far below it.

That's why a small loss can feel intolerable, and a large loss can feel weirdly "already gone," producing the dangerous thought: *might as well hold—it can't get worse*. It can.

Now add a second force: regret. Regret is not just sadness about a bad outcome. It's what happens when your mind replays the alternative movie—what you *could* have done—and punishes you for not choosing it. Regret is uniquely corrosive in markets because markets generate an infinite number of alternative movies. Every tick creates a new universe where you bought earlier, sold earlier, rotated, hedged, or "listened to your gut."

Regret doesn't require you to be wrong. It only requires that another path existed. You can make a sensible decision, get unlucky, and still feel regret—because your brain judges the decision by the outcome, not by the logic that produced it.

Loss aversion and regret combine into a particularly nasty cocktail: *the fear of being wrong in public.* A loss is painful; a loss that you feel responsible for is worse; a loss that you could have avoided (in your imagination) is worst. This is why investors often prefer the safety of "consensus mistakes." If everyone was doing it, the regret feels diluted. If you were the lone contrarian and you're wrong, the regret feels concentrated—like a spotlight.

This is also why so many destructive behaviors come wrapped in the language of responsibility. "I'm just managing risk." "I'm protecting gains." "I'm being prudent." Often, those are true. Sometimes they're just socially acceptable disguises for a simpler motive: *I can't stand the feeling of this going lower with me watching.*

Prospect Theory makes another prediction that sounds like a paradox until you've lived it: people tend to be risk-averse in gains and risk-seeking in losses. When you're up, you want to "lock it in." When you're down, you want to "make it back."

The market turns that into a behavioral seesaw:

- When you're winning, you sell too early to avoid the pain of giving back gains.
- When you're losing, you hold too long (or even double down) to avoid crystallizing the loss and admitting you were wrong.

That pattern has a name in finance research—the disposition effect—but you don't need the jargon to recognize the feeling. A winning position feels like a fragile gift you must not drop. A losing position feels like a debt you must repay. Neither framing is a calm assessment of expected future returns. Both are emotional accounting.

There's a specific symptom worth diagnosing because it shows up everywhere from meme stocks to perfectly respectable blue chips: *get-even-itis*. It sounds minor, like a cold. It is not. It's a mental model that says: "My job is to get back to zero." Not "my job is to allocate capital to the best opportunities from here." Not "my job is to reach my goals." Just: zero.

Zero becomes sacred because it's the point where the loss stops hurting and the regret stops accusing. Zero is psychological amnesty.

This is how people end up holding a mediocre investment for years because selling at a loss would "make it real." But the loss is already real. The only thing unreal is the hope that the market owes you a return path back to your entry price. It does not know your entry price exists.

If you want a quick test for get-even-itis, ask yourself: *If I didn't already own this, would I buy it today?* If the honest answer is no, the position is no longer an investment thesis; it's an emotional negotiation.

The same dynamics infect "taking profits." Selling winners can be wise—sometimes a position has become too large, the valuation has changed, or the risk has increased. But many profit-takes are not analysis; they're relief. The investor is paying an invisible fee to purchase the feeling of being right.

Relief is expensive when it repeatedly interrupts compounding.

Another Prospect Theory concept is the certainty effect: people overweight outcomes that feel certain relative to outcomes that are merely probable. That's why a guaranteed small gain can feel more attractive than a larger expected gain with uncertainty attached. Translating into investing behavior: after a scary drawdown, a money market fund yielding something "sure" can feel like oxygen—even if, over decades, it's not enough to reach the goal.

The emotional logic is powerful: *I can't take another hit. Give me something safe.* The math is equally powerful: over long horizons, safety that undershoots your required return is its own kind of risk—the slow-motion risk of not getting where you need to go.

This is where framing becomes a lever you can actually use. A 30% decline in a diversified equity portfolio is framed as "losing almost a third" if your reference point is last month's peak. It can also be framed as "a temporary markdown" if your horizon is decades and your plan assumed drawdowns would happen. Both frames describe the same reality. The market didn't change. Your interpretation did.

And interpretation is not mere positive thinking; it determines whether you trigger costly actions. A drawdown that is framed as "danger" invites selling. A drawdown framed as "expected turbulence" invites rebalancing, continuing contributions, or simply staying still.

Right next to the explanation of why a 1,000*losshurtsmorethana*1,000 gain feels good, it helps to see the shape that drives all of this:

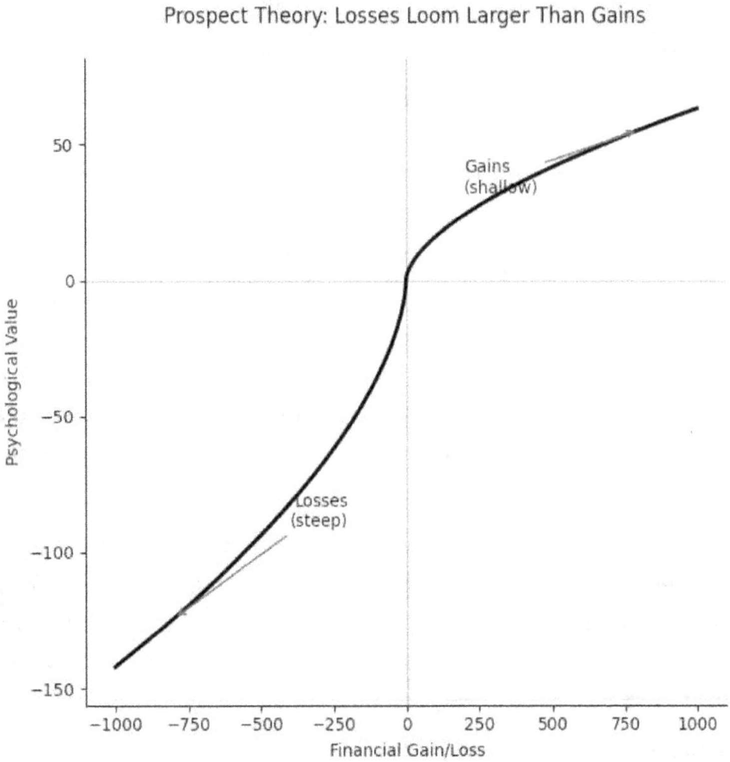


Once you've internalized that curve, a lot of "mysterious" investor behavior becomes painfully logical.

Consider the investor who sells after a drop because they "can't risk losing more." That decision is rarely the result of a new calculation about long-term expected returns. It's the desire to stop the pain on the left side of the curve. Selling converts an uncertain future into a certain present: the loss is finalized, and the anxiety stops spiking every time the market opens.

The cost is subtle and brutal: you may have just traded a temporary drawdown (a feature of equity ownership) for a permanent reduction in future compounding. The pain ended, but the plan got poorer.

Now consider the investor who refuses to sell a losing single stock because "it'll come back." Sometimes it will. Sometimes it won't. The important part is *why* the thought is so sticky. Selling locks in the loss and invites regret: "I chose this. I was wrong. I should have known." Holding keeps the alternate movie alive: "In the other version of the future, I'm vindicated." The position becomes less about expected value and more about emotional redemption.

A portfolio can quietly turn into a museum of unresolved feelings.

Regret is also why people chase. Buying something after it has already soared is often framed as greed, but there's a more human explanation: the investor is trying to avoid the regret of watching it go even higher without them. The purchase is not an investment; it's a regret hedge.

Unfortunately, regret hedges tend to be purchased at the worst possible time: when the story is most seductive, the crowd most confident, and the price least forgiving.

To make this concrete, here's a small table of common "thoughts" and what they usually indicate. The point isn't to shame yourself; it's to recognize a pattern early, before it turns into a trade you'll later rationalize.

| *Thought* | *What it often signals* |
|---|---|
| "I'll sell when it gets back to what I paid." | Reference-point thinking; get-even-itis; ignoring opportunity cost. |
| "I should take profits before they disappear." | Loss aversion applied to gains; preference for relief over expected value. |
| "I can't take another down day." | Pain minimization; converting uncertainty into certainty (often at the wrong time). |
| "Everyone else is making money in this." | Regret avoidance + social proof; buying to avoid feeling left out. |

Notice what's missing from the left column: anything about the asset's future cash flows, competitive position, or role in a diversified plan. That omission is the tell. When the internal monologue is dominated by reference points and feelings, the decision is being made on the psychological axis, not the financial one.

There's also a counterintuitive twist: loss aversion can make investors *more* reckless. When you're down, the curve predicts risk-seeking behavior because a gamble that might restore you to the reference point feels emotionally attractive. This is how "I'll just wait" becomes "I'll double down," and how "I'll diversify later" becomes "I need one big win."

The market doesn't require you to become a gambler. It only needs you to become desperate.

A good long-duration plan treats desperation as a risk factor, like leverage. Desperation is a kind of internal margin loan: it tempts you to magnify risk at the exact moment your judgment is compromised.

So what can you do with all of this besides nod grimly?

Start with a simple but powerful reframe: the goal is not to avoid losses. The goal is to avoid *behavioral* losses—actions taken to escape discomfort that permanently damage the plan. Market losses are sometimes unavoidable. Behavioral losses are optional, but only if you design for them.

The second step is to relocate your reference point. If your reference point is last month's account high, you will live in a permanent state of disappointment because markets regularly fall below recent highs. A more useful reference point is your written plan: your target allocation, your time horizon, and your required return. The plan should become the "normal," not the latest peak.

That sounds abstract until you do one practical thing: stop tracking performance against your purchase price for long-term holdings. Purchase price is emotionally loud and often financially irrelevant. It tempts you to treat markets like a courtroom—innocent until proven guilty—rather than like a capital allocation problem.

If you need a number to anchor to, anchor to *process*:

- "Am I still diversified the way I intended?"
- "Am I still saving at the rate that makes the plan work?"
- "Am I rebalancing when allocations drift?"

Those questions are less thrilling than "Am I up today?" That's why they're effective.

Finally, treat regret like a weather report, not a command. Regret will show up no matter what you do, because markets create alternate movies endlessly. Your job is not to eliminate regret. Your job is to stop paying it protection money.

*Actionable takeaway:* Write a one-page "regret script" and keep it where you place trades. It should contain (1) your long-term horizon in years, (2) your target asset allocation, and (3) a pre-commitment sentence you will follow when you're down, such as:

> "When I feel the urge to sell to stop the pain, I will wait 48 hours, reread my allocation plan, and only act if I can explain the decision without mentioning my purchase price or recent highs."

The point isn't to be heroic. It's to make sure that if you do act, you're acting from your plan—not from the steep left side of the curve.

## 4.3 Ego, Luck, and Overconfidence

The most seductive sentence in investing is not "this can't go down." It's "I was right."

A rising market can make almost anyone look competent. Buy a handful of well-known stocks during a broad rally, and your account starts producing little green affirmations. The brain reads those affirmations the way it reads applause: not as *market weather*, but as *personal merit*. After a while, the story shifts from "I got lucky" to "I've got a feel for this." And that's when the compounding machine quietly acquires a new moving part: ego.

Ego is not just vanity; it's a decision-making filter. It edits your memory, polishes your narrative, and turns uncertainty into something that feels like skill. In a bull market, ego is constantly fed. You don't simply make money-you make money *while holding opinions*. Your brain loves that combination. It's one thing to be paid. It's another to be paid *and vindicated*.

That's the opening where overconfidence walks in wearing a lab coat.

Overconfidence, in investing, is the belief that your judgment is more accurate than it really is. Sometimes it shows up as "I can pick winners." Sometimes it shows up as "I can time exits." Often it shows up as "I can handle more risk than other people." It's rarely framed as arrogance, because it doesn't feel like arrogance from the inside. It feels like realism. You're not imagining you're invincible; you're noticing you've been right lately. The problem is that "lately" is a terrible dataset.

If you want to see the mechanics, watch what happens after a few wins in a row. The first win is fun. The second win feels like confirmation. By the third, you start to attribute the outcome to a stable trait: insight, discipline, the ability to spot patterns, the courage to act when others hesitate. You begin to generalize. The mind hates randomness, and a streak of wins is the raw material for a story about who you are.

Then the story demands a bigger stage.

Overconfidence doesn't merely increase risk; it concentrates it. A diversified portfolio is emotionally unsatisfying because it refuses to flatter you. If you're right about the market's direction, a diversified portfolio says, "Nice, but so was everyone else." If you're right about a single stock or a narrow theme, it says, "You saw what others missed." Concentration is not just a financial bet-it's an identity bet. It's you, distilled.

That's why concentration so often follows success. It's not that people wake up wanting fragility. They wake up wanting meaning.

You can see it in the phrases that start to appear in the self-talk:

> "This is my highest-conviction idea."
> "I know this business better than the market does."
> "Diversification is for people who don't do the work."
> "I'm not trading—I'm *investing*."

Some of these sentences can be true in rare cases. The danger is how easy they are to say when the market has been kind.

The finance world has a polite term for this: the illusion of control. The idea that because you are active—researching, watching, clicking, adjusting—you are increasing your odds in a domain where outcomes are heavily influenced by variables you can't see and can't control. Activity feels responsible. Stillness feels lazy. The market happily exploits that intuition by offering endless levers.

A brokerage app is basically a cockpit full of buttons attached to a weather system.

The proof that this isn't just a philosophical complaint is sitting in the data: individual investors who trade more tend to earn less, net of costs. That's not a moral judgement about effort; it's a statement about frictions. Every trade has a spread, potential market impact, taxes, and the chance you're trading against someone with more information or faster execution. Overconfidence doesn't just tempt you to take more risk; it tempts you to pay more tolls.

And the tolls are sneaky because they arrive as tiny cuts. A few basis points here, a few short-term gains taxes there, a few timing errors that feel "unlucky." Death by a thousand paper cuts is still death.

There's also a social version of overconfidence that is almost impossible to resist: the "above-average" belief. Most drivers think they're above-average drivers. Most investors think they're above-average investors. The math is not friendly to that belief. Markets are competitive. Before costs, the average dollar earns the market return. After costs, the average dollar earns less. To be above average, someone has to be below average—and the market does not send a courtesy email telling you which one you are.

Overconfidence is easiest to spot in other people and hardest to spot in yourself, because it often looks like a virtue: conviction.

Conviction is not automatically foolish. The world has been improved by people with conviction. Investing, however, charges a specific kind of penalty for misplaced conviction: it turns your *confidence* into your *position size.* It converts a psychological state into a financial outcome. You can be wrong in a small position and recover easily. You can be wrong in a concentrated portfolio and spend years digging out.

That's why concentration risk is the most important practical consequence of overconfidence. A concentrated bet can make you rich. It can also ruin the compounding engine at the worst possible time: not by a 10% bruise, but by a 60% fracture that takes years of returns just to return to even.

Even worse: the bigger the win, the stronger the incentive to concentrate further. Imagine a single stock becomes the hero of your portfolio. You didn't just make money—you made money *in a way that proves you're smart*. Selling feels like heresy. Diversifying away feels like admitting the win was luck. The holding becomes sacred.

Sacred holdings are where portfolios go to become fragile.

Sometimes the concentrated position is a "story" stock. Sometimes it's crypto. Sometimes it's a sector ETF that rode a big wave. And often, it's the most psychologically loaded asset of all: your employer's stock. Employer stock concentration is overconfidence with a uniform on. You know the products. You know the leadership. You know the internal jargon. Familiarity feels like insight.

Familiarity is not diversification.

The harsh irony is that employer stock ties together multiple risks you don't want tied together: your salary, your benefits, and your portfolio. If the company hits trouble, you can lose income at the same time your concentrated holding declines. It's not just concentration; it's correlation—bad things happening together.

Overconfidence also has a "timing" flavor. After a few good calls, you start to believe you can step out before the fall and step back in after the dust clears. Timing is attractive because it promises a magical combination: the upside of risk assets without the pain of drawdowns. It's the behavioral equivalent of wanting to surf without getting wet.

The problem is that market timing asks you to be right twice: when you sell and when you buy back. The first decision is often driven by fear; the second is often blocked by pride. If you sell and the market rises, you don't want to buy higher because that would confess a mistake. So you wait for a pullback that may not arrive. The longer you wait, the more expensive your pride becomes.

Overconfidence doesn't merely lead to bad decisions. It can trap you in them.

There's a particularly cruel loop that shows up after a win:

1. You take a risk.
2. The market rewards you (perhaps with a large gain).
3. You infer skill.
4. You increase risk.

The loop is self-reinforcing right up until it isn't. This is how people "graduate" from index funds to options. It's how a few lucky trades become a belief system. It's how a portfolio goes from resilient to brittle while the owner feels more and more confident.

The market loves brittle confidence because it tends to break at the same time: when volatility returns.

A useful mental model here is to separate *outcome quality* from *decision quality*. A decision can be good and still lose money. A decision can be bad and still make money. Investing is one of the few domains where bad decisions can be rewarded long enough to teach the wrong lesson.

That's why bull markets are dangerous teachers. They grade on a curve, and the curve is rising.

If you want a more honest teacher, look at your process rather than your profits. Did you size positions according to a written plan?

Did you diversify across drivers of return? Did you avoid leverage you couldn't survive? Did you rebalance rather than chase? Did you account for taxes and costs? Those questions don't give you the dopamine hit of a green day, but they are far more predictive of whether your wealth machine will still be running when the market gets ugly.

Overconfidence hates process because process refuses to applaud.

It helps to name the "overconfidence signals" that show up in normal language. Here are a few that tend to matter:

- *"I don't need diversification; I understand this one."* (translation: you're confusing familiarity with risk control)
- *"I'll just watch it closely."* (translation: you think attention reduces volatility)
- *"I have a high risk tolerance."* (translation: you haven't tested it in a real drawdown with real dollars)
- *"I can always get back in."* (translation: you're underestimating the psychological difficulty of buying after you sold)

These aren't condemnations. They're early-warning indicators. The earlier you catch them, the less tuition you pay.

There's also a quiet demographic wrinkle that shows up in studies of trading behavior: on average, men trade more than women, and that extra trading has historically been associated with lower net returns. The takeaway isn't that one gender is "better" at investing by nature. The takeaway is that confidence can be costly when it turns into activity, and activity is one of the easiest ways to leak returns without noticing. The market doesn't care why you traded. It only cares that you paid to do it.

Overconfidence is rarely cured by reading. It's cured by *constraints*.

Constraints are not limitations; they're safety rails. They keep your future self—who may be euphoric, angry, embarrassed, or convinced they've "cracked it"—from making a decision that permanently damages your plan.

One of the cleanest constraints is position sizing with a hard ceiling. Decide in advance what "too big" means for any single stock, sector, or theme. For many long-term investors, a sensible ceiling for a single stock might be 5% or 10% of the portfolio; a ceiling for a single sector might be something similar, depending on the strategy. The exact number is less important than having a number before you're emotionally attached.

A second constraint is a rebalancing rule that forces humility. Rebalancing—selling a bit of what has gone up and buying a bit of what has gone down—feels like the opposite of what your ego wants. Ego wants to ride the winner and abandon the loser. Rebalancing says: *the winner is now a bigger risk; trim it*. It turns success into a prompt for discipline rather than a prompt for escalation.

A third constraint is a "trade tax" you impose on yourself: every discretionary trade must clear a written hurdle. Not "I feel like it," not "Twitter says," not "it's breaking out," but a short written thesis and a pre-defined reason you would admit you're wrong. Writing sounds boring. That's the point. Boring filters out impulsive trades the way a bouncer filters out trouble.

And if you're tempted to argue that you're the exception—good. That temptation is the signal.

*Actionable takeaway:* Write an "ego cap" rule and implement it mechanically. Choose one of the following and commit in writing:

1. A maximum allocation to any single stock (e.g., 10%) and a quarterly trim back to the cap, or
2. A maximum number of discretionary trades per month (e.g., 2), with unused trades not rolling over.

Then add one sentence beneath it: *"When I feel most certain, I will treat that feeling as a reason to reduce risk, not increase it."* That single reversal—certainty as a warning sign—can keep a lucky streak from turning into a permanent compounding injury.

## 4.4 Narratives, Herding, and FOMO

The most dangerous investment research is the kind that makes you feel like you've joined a team.

It starts innocently. A friend mentions a stock at dinner. A coworker says, "I can't believe how much this is up." A podcast host laughs and calls it "the easiest trade in the world." You don't feel greedy yet. You feel *late*. And "late" is a social emotion before it's a financial one—like arriving to a party where everyone already has a drink and an inside joke.

FOMO (Fear Of Missing Out) is often described as childish impatience. That misses the point. FOMO is a sophisticated survival instinct in modern clothing. For most of human history, being out of sync with the group was dangerous. If everyone ran, you ran. If everyone avoided a certain area, you avoided it. Social alignment wasn't a personality trait; it was a risk-control system.

Markets are one of the few places where the tribe can sprint straight toward a cliff—and your brain still screams that the real danger is being left behind.

What makes this especially tricky is that the crowd is not always wrong. Sometimes the herd moves because it has real information. Sometimes a story is true. Sometimes a technology really does change everything. The trap is that the brain tends to skip the hard part—*price*—and focus on the easy part—*narrative*. When the narrative is compelling, we treat it like a substitute for valuation, risk, and time horizon.

A story is emotionally complete. A stock is not.

A narrative has a hero, a villain, and a moral. It has a clean arc. It has an ending you can imagine. Investing rarely offers that. Investing is a foggy forecast plus a probability distribution plus a lot of waiting. So the mind does what it always does when it hates uncertainty: it grabs a story that feels like certainty and treats it as a map.

Once a narrative takes hold, herding follows naturally—not because people are stupid, but because people are social. You don't need a formal conspiracy for a bubble. You need a shared storyline, a rising price chart, and a crowd of humans who would prefer to be wrong together than wrong alone.

There's an underrated emotional benefit to the herd: it spreads blame. If you buy what everyone is buying and it fails, you can say, "Nobody saw this coming." If you buy something unpopular and it fails, you are forced to say, "*I* saw it wrong." Regret is personal pain, and the crowd is an anesthetic.

This is why "everyone's doing it" feels like safety, even though it's often the opposite.

Herding has a technical-sounding cousin: information cascades. The idea is simple and unsettling. In a cascade, people stop relying on their private information and instead copy what others appear to be doing. Not because copying is rational in some abstract sense, but because other people's actions contain information—or at least they *seem* to.

If you see enough people buying, you infer they must know something. Then your buying becomes the evidence the next person uses. The feedback loop can become self-sustaining. The market turns into a giant game of "follow the leader," except the leader might be a rumor, a charismatic influencer, or a short-term price move that was never meaningful in the first place.

In an information cascade, price becomes proof. And proof feels like permission.

This is the psychological core of bubbles: not irrationality, but the substitution of social evidence for fundamental evidence. You stop asking, "What is this worth?" and start asking, "How high can it go?"

That question is not foolish—it's a different game. It's the game of musical chairs. It can be profitable for a while. It becomes ruinous the moment the music stops, because the exit is crowded and the chairs were never real.

The best bubbles are built on narratives that are *partly true*. The dot-com era didn't hallucinate the internet. The internet really did rewire the economy. The problem was that investors correctly identified a revolution and then paid any price for a ticket to it. They bought the future and forgot they were still buying *claims on cash flows*. The narrative was right; the math was wrong.

That pattern repeats because it's emotionally satisfying. A grand story offers moral clarity: this is the new world; those who doubt will be left behind. Skepticism becomes not just a financial stance but a social faux pas. You're not cautious—you're "out of touch." You're not prudent—you "don't get it." It's hard to stay disciplined when discipline is framed as ignorance.

FOMO is not only the fear of missing returns. It's the fear of missing *membership*.

Recency bias pours gasoline on the fire. Recency bias is the tendency to extrapolate the immediate past into the future. If something has been going up, we expect it to keep going up. If it has been going down, we expect it to keep going down. In markets, that's a particularly expensive habit because the most dramatic moves are often followed by reversals. The chart that feels like a reliable trend is sometimes just a stretched rubber band.

The brain, however, is not built to feel rubber bands. It's built to feel trajectories.

A key detail: recency bias doesn't usually announce itself as "bias." It announces itself as *common sense*. "Look at the chart." "This is working." "The world has changed." "Institutions are buying." The evidence is real—prices really have moved. The mistake is assuming the move makes the asset safer. Often, the move makes the asset *riskier*, because a higher price means a lower future return unless the fundamentals improve even faster than the price did.

Rising prices create comfort. Comfort creates buying. Buying creates rising prices. It is an emotional perpetual motion machine—until the energy runs out.

One reason standing apart from the crowd feels physically painful is that social pain and physical pain share circuitry. Rejection, mockery, and exclusion are processed in the brain in ways that resemble bodily threat. So when you refuse to join the mania everyone is celebrating, you're not just making a financial decision; you're risking social discomfort. You are choosing to be the person who "doesn't get it," the person who "missed it," the person who might be proven wrong by a chart that keeps climbing.

It's easy to admire contrarianism in hindsight. In real time, it feels like swallowing needles.

That's why the most reliable contrarians are rarely "brave" in the cinematic sense. They are procedural. They don't resist the crowd by force of personality; they resist by force of rules. They have an allocation target. They have rebalancing bands. They have limits. They have an identity that is not built around being right in public.

Because if your identity is "the smart one," you are highly vulnerable to narratives. Narratives are social currency. They offer the chance to be early, to be in-the-know, to have a hot take that becomes a victory lap. The market doesn't pay for hot takes. It pays for owning assets with durable cash flows at reasonable prices and holding them through discomfort. Unfortunately, hot takes are much more fun to talk about at dinner.

The market's most expensive product is entertainment.

There's a special kind of narrative that deserves its own warning label: the moral narrative. This is when an investment becomes a stand for virtue. "This company is changing the world." "This technology is liberating people." "This movement is inevitable." Moral narratives are powerful because they recruit your identity. If you question the investment, you are not just questioning expected returns—you are questioning goodness, progress, or belonging.

Once an investment becomes moralized, selling feels like betrayal, and skepticism feels like cynicism. That's how people hold through obviously deteriorating conditions: they are not holding a stock; they are holding a worldview.

A quieter narrative trap is the "smart money" story. You see that a famous investor, fund, or institution owns something, and you assume their ownership validates the trade. It might. But remember: large institutions have different time horizons, different risk constraints, different hedges, and different entry prices. You are seeing a headline. They are seeing a portfolio.

Copying "smart money" without context is like copying a surgeon's hand movements without knowing the diagnosis.

Now, if narratives and herding are so powerful, why doesn't everyone get swept up every time? Because many people do have private information: not about the market, but about themselves. They know their goals, their time horizon, their need for liquidity, their tolerance for drawdowns, and the role their portfolio plays in their life. That self-knowledge is the most important "edge" most investors will ever have. It's also the first thing FOMO tries to steal.

FOMO doesn't ask, "Does this fit your plan?" It asks, "Can you stand being the only one not getting rich?"

There's a practical way to protect yourself: force narratives to compete against numbers. Not complicated numbers. Just the few that matter.

When a story grabs you, ask two blunt questions:

1. *If this goes down 50%, will I still be okay—financially and emotionally?*
2. *If this doubles from here and I miss it, will my plan still work?*

Those questions convert FOMO from a vague ache into a concrete trade-off. Most of the time, the answers are revealing. You realize the "must buy" is actually optional, and the "can't miss" is mostly ego.

And then there's the most counterintuitive lesson of all: sometimes the crowd is right, and you still shouldn't join. Not because you're smarter, but because you're running a different race. A 22-year-old with a small account can take risks that a 55-year-old nearing retirement cannot. A professional with stable income can take risks a freelancer can't. A person with strong savings habits can recover from mistakes that a chronic spender can't. The market is one arena; lives are many.

You don't need to win *their* game. You need to win *yours*.

That's the difference between investing and social performance. Investing is allowed to be boring. Social performance demands novelty. Investing rewards patience. Social performance rewards being first.

Your portfolio does not care whether you were first. It cares whether you were solvent and disciplined.

*Actionable takeaway:* Create a "narrative quarantine" rule. Any time you feel a strong urge to buy because of a story—especially a story you want to repeat to other people—you must wait 72 hours and write a one-paragraph memo answering:

> "What would have to be true for this to be a bad investment *at today's price?"*

If you can't write that paragraph, you don't have an investment thesis; you have a membership application. The waiting period won't kill good opportunities, but it will kill a surprising number of crowd-chasing impulses—and that alone can save years of compounding.

## 4.5 The Cost of Capitulation

The market has an unadvertised fee schedule. It's not the expense ratio, not the advisory fee, not the bid—ask spread. It's the fee you pay when your stomach takes over your mouse hand. It shows up as a sell button pressed at the exact wrong time, a buy order placed in a euphoric haze, a portfolio "clean-up" that feels responsible and ages like milk.

It's a fee with a cruel accounting rule: you don't get an invoice. You get a quieter future.

That fee is the *behavioral gap*—the distance between what markets return and what investors actually capture. It's not a moral failing. It's math plus biology. The math says compounding depends on staying in the game. The biology says volatility feels like danger, narratives feel like certainty, and regret feels like punishment. Put those together and you get the most common way long-term wealth plans fail: people don't blow up once; they leak returns repeatedly.

The behavioral gap has a simple mechanical cause that hides behind complex psychology: money tends to arrive *after* good returns and leave *after* bad returns. That timing difference matters because markets don't compound your opinions. They compound your dollars. If your dollars are invested during the dull middle and sitting in cash during the violent rebound, your personal return becomes a shadow of the market's return—even if you owned the same funds at some point along the way.

This is why "I owned the S&P 500 for years" can be true and still not produce anything like the S&P 500's long-term results. Ownership is not a binary state. It's a timeline. Your timeline is your return.

To make this less abstract, it helps to separate two kinds of returns that sound similar but behave very differently:

- *Market (or fund) return*: what the investment itself earned over a period.

- *Investor return*: what *you* earned, given when you added money, when you pulled it out, and when you switched.

The second one is the one that pays for your retirement. It's also the one most investors unintentionally sabotage.

The sabotage is usually not a dramatic day-trading montage. It's a very normal human pattern: *capitulation*. Capitulation is the moment you stop trying to be "right" and start trying to stop hurting. The news is terrifying, the portfolio is red, your group chat has turned into gallows humor, and you decide you can't take another day of this. So you sell. Or you "go to the sidelines." Or you "wait for clarity."

Clarity, in markets, usually arrives after prices have already moved.

Capitulation doesn't feel like greed or fear in the cartoonish sense. It feels like relief. That's why it's so dangerous. Relief is an emotion you can buy with a trade. The price is often a permanently lower base for compounding.

Here's the part that feels like a prank: the best days in the market often cluster around the worst days. Volatility isn't a smooth wave; it's a storm system. The big drops happen during crisis. The big rebounds also happen during crisis. If you sell in the middle because you can't stand it, you frequently miss the rebound—the exact set of days that do the heavy lifting for long-run returns.

Missing a handful of those days is not a small "oops." Over multi-decade horizons, it can be catastrophic.

The reason is multiplicative. Compounding is a chain, and missing a strong link doesn't just reduce one year's return; it reduces the base on which all future returns build. A mistake in year 3 echoes in year 20. The market is forgiving about many things. It is not forgiving about interruptions to compounding.

It's worth pausing on why "best days" matter so much. Returns are not evenly distributed. A surprisingly large fraction of long-run gains can come from a relatively small number of outlier days. That's uncomfortable because it denies the investor's favorite fantasy: "I'll just be out during the bad parts and in during the good parts." The market doesn't label them in advance. Often the "good parts" happen while headlines are still screaming.

The market's biggest gifts arrive in ugly wrapping.

To put numbers on the intuition, consider a stylized 20-year example: start with $10,000 in an equity index and let it ride. Now compare that to what happens if you miss just the 10 best days, or the 20 best days, over the same period. The exact results vary by the start and end dates, but the direction is stubbornly consistent: missing a small number of the best days can cut your ending wealth dramatically.

And you don't miss the best days because you were on vacation without Wi-Fi. You miss them because you were in cash, hiding from the storm you could no longer stand.

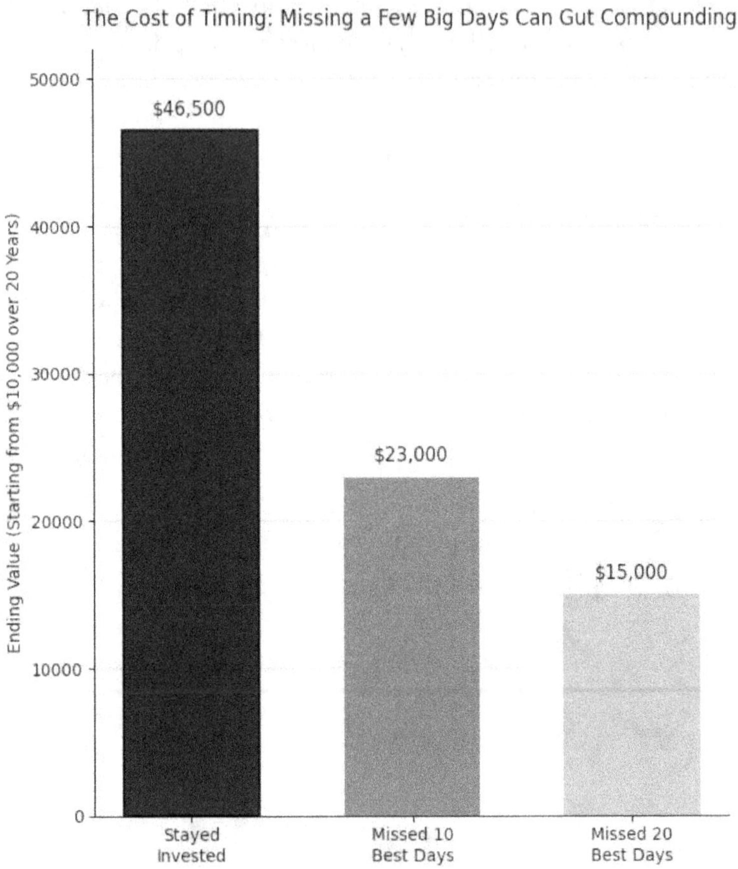


Illustrative example: exact results depend on the specific 20-year window, but the asymmetry is a persistent feature of equity returns.

The important part of that chart isn't the exact dollar amounts; it's the shape. The "stayed invested" bar towers over the others because outlier days matter, and outlier days are easy to miss if you treat investing as something you can "step away from" during stress.

This is the hidden irony: many investors sell because they want to reduce risk, but by stepping out they may increase the most dangerous risk of all—the risk of not being there when the market does what it historically does after panic: rebound.

That doesn't mean the market always rebounds quickly. It doesn't. Sometimes the grind back takes years. But the *sharp* rebound days—the ones you can't easily replicate—tend to show up when sentiment is still awful. Those are exactly the days you're least likely to be invested if your strategy depends on emotional comfort.

So the behavioral gap is not just "people buy high and sell low." That phrase is true but too polite. The deeper truth is that investors often sell low *and then fail to buy back* because buying back is psychologically hard. After you sell in a panic, you don't feel calm; you feel fragile. The market rising without you feels like punishment. Re-entering feels like admitting you were wrong. So you wait for "a better entry."

The better entry often never comes.

Or it comes, briefly, and you don't take it because now you're scared of being hurt again. Your brain, having been burned, treats re-entry as a fresh source of regret. This is the compounding damage: one capitulation trade can create months or years of underexposure, and underexposure is a silent killer when your plan requires equity-like returns.

It's also why the investor return gap can exist even in broadly diversified funds. You can choose a perfectly sensible allocation and still sabotage it by changing it at the wrong times. Asset allocation is the engine. Behavior is the steering wheel. Engines don't help if the steering wheel keeps yanking toward the guardrail.

The behavioral gap has a cousin that's equally common: performance chasing. When markets have been strong, investors feel safe and optimistic. That's when they add money, switch into the hottest fund, increase equity exposure, or abandon diversification for whatever has the best recent chart. They are not thinking, "I will buy high." They are thinking, "The world is improving. This is the new normal."

Then volatility returns, and the same investors reverse the trade. They don't just lose money in the drawdown; they miss the rebound because they're still licking their wounds. It's a two-hit combo: buy late, sell early.

At scale, that cycle turns the market into a treadmill. Lots of sweating, not much distance.

There's another subtle mathematical trap: the emotional narrative of "getting back to even." A person who capitulates after a big drop often frames the future as a chance to recover what was lost. But if they're in cash while the market rises, they are not recovering; they are watching recovery happen to someone else. They become trapped in a resentment trade: "I'll get back in when it's fair." Markets have no fairness department.

Compounding is not a courtroom. It's a machine. You either stay connected to it, or you don't.

None of this is meant to imply you should never sell anything, ever. Sometimes selling is correct: if you need the money soon, if your risk capacity changes, if your portfolio was mis-sized, if you were speculating rather than investing, if you were concentrated in something that no longer fits your plan. Selling can be prudent.

The behavioral gap comes from selling for a different reason: selling because you can't tolerate your own emotions *and* you don't have a plan for how you'll re-enter.

Capitulation is not just an exit. It's an exit without a map back in.

So how do you build a practical defense? Start by treating "staying invested" as an operational task, not a personality trait. You don't rely on being brave in the moment. You rely on having pre-decided rules that do the brave thing on your behalf.

A useful way to do this is to create an explicit "volatility plan." The plan answers two uncomfortable questions in advance:

- *What will I do if my portfolio drops 20%?*
- *What will I do if it drops 40%?*

If your honest answer is "I don't know," your strategy is not a strategy. It's a mood.

A second defense is to reframe rebalancing and contributions as your anti-capitulation machinery. When prices fall, your target allocation becomes your north star. If equities are down and your plan says you hold equities, the plan has a built-in counterpunch: rebalancing forces you to buy what got cheaper. Contributions do the same thing automatically. They convert scary headlines into a mechanical purchase schedule.

This is one of the few times in finance where "set it and forget it" is not laziness; it's wisdom.

A third defense is to reduce the frequency with which you measure your portfolio. Measurement creates emotion; emotion creates action; action creates fees and mistakes. For a long-horizon investor, daily checking is like pulling a plant out of the soil each morning to see if it grew. You don't get better growth. You just damage the roots.

And there's a final defense that sounds almost too simple: write down what "capitulation" looks like for you personally. For one person it's doomscrolling. For another it's calling a friend who loves drama. For another it's making "small" trades that add up. Your own warning signs are the earliest point at which you can intervene.

*Actionable takeaway:* Draft a one-page "Drawdown Playbook" and store it somewhere you will actually see when markets are ugly (printed in a folder, saved as a phone note, taped inside a notebook). It must include:

1. A rule that forbids panic selling (e.g., "No equity sales within 7 days of a market drop of 5%+ unless cash is needed within 12 months.")
2. A pre-committed action that turns volatility into process (e.g., "If equities fall 20% from peak, rebalance back to target allocation within 10 trading days.")
3. A re-entry clause if you do reduce risk (e.g., "If I move to cash, I will schedule equal buys over the next 6 months regardless of headlines.")

The playbook won't prevent drawdowns. It will prevent the more expensive thing: converting a temporary drawdown into a permanent compounding wound.

## 4.6 Behavioral Systems and Guardrails

Willpower is a terrible asset class.

It performs brilliantly in backtests and collapses in live trading. It looks strong on a quiet Sunday afternoon and turns illiquid the moment the market gaps down, your phone lights up with headlines, and your group chat starts posting skull emojis in all caps. If your investment plan depends on "I'll stay calm," your plan is not a plan; it's a mood with a brokerage account.

The practical answer is not to become braver. It's to become more automatic.

The market is a pressure test that arrives on a schedule you don't control. So the only reliable defense is to build a system that assumes your future self will be stressed, distracted, and fully capable of making an expensive decision for reasons that will sound "rational" in the moment. The goal is not to remove emotion; it's to keep emotion from rewriting your portfolio.

A good behavioral guardrail does one of two things:

- It makes the right action easier than the wrong action.
- It makes the wrong action slower than your impulse.

Ideally, it does both.

This is why retirement plans work better than retail accounts for many people, even when the available funds are similar. The secret sauce isn't access to fancy products. It's the architecture: automatic contributions, default investments, friction around withdrawals, and the quiet social norm that you're "supposed" to leave it alone. Inertia—normally blamed for laziness—gets harnessed as a compounding engine.

In other words, the best investing behavior is often not discipline. It's *default*.

You can borrow those design principles and install them in your own financial life. Think like an engineer, not a monk. Engineers don't demand perfect behavior from unreliable components. They design around them.

The flagship tool for doing this is an Investment Policy Statement (IPS). The name sounds like something written by a committee in a windowless room. Good. You want that vibe. You want a document with the emotional temperature of a refrigerator—because it will eventually have to overrule a brain that is on fire.

An IPS is a short written contract between *you now* and *you later*. "You later" is the version of you who will feel certain right before being wrong, fearful right before a rebound, and tempted to chase the hottest thing right after it's already hot. "You now" is the version of you capable of admitting that.

The IPS does not predict the future. It prevents you from improvising one.

Most investors resist writing an IPS for the same reason they resist wearing a seatbelt on a short drive: it feels unnecessary right up until it isn't. But the entire point is that it will be used during the handful of moments when your instincts are most likely to betray you. Those moments are rare. They're also decisive.

A practical IPS is not long. If it's too long, you won't read it when you need it. One to two pages is enough. The content should be specific enough to constrain behavior and flexible enough to survive real life.

Here's what makes an IPS actually functional rather than ceremonial:

### 1) Define the mission in plain language

The mission is not "maximize returns." That's a gambling slogan. A real mission is tied to the life the portfolio must fund: retirement income, a future home, financial independence, charitable goals, legacy, or simply not being forced into bad choices later.

The mission statement should be boring and undeniable, like:

> "This portfolio exists to fund long-term goals over 15+ years. I will accept volatility as the price of higher expected returns."

The line about volatility is not poetry; it's a psychological inoculation. You are reminding yourself that the pain is not evidence of failure. It's evidence that you are holding risk assets.

### 2) Specify the allocation---and what counts as ``off plan"

Asset allocation is just a fancy phrase for "what percentage goes where." It's the most important lever most investors can pull because it determines how much volatility you will experience and how much expected return you can plausibly earn.

A useful IPS includes:

- Target allocations (e.g., 70% stocks, 30% bonds).

- Ranges that define drift (e.g., stocks can float between 65%–75%).
- The instruments you'll use (broad index funds, specific ETFs, etc.).

The ranges matter because they create a trigger. Without a trigger, rebalancing becomes a debate. Debates are exactly what you lose during stress.

When the market is calm, you can debate. When the market is panicking, you need a switch that clicks.

### 3) Pre-commit to rebalancing

Rebalancing is the least glamorous form of courage. It forces you to trim what has gone up (when it feels safe) and add to what has gone down (when it feels dangerous). It turns volatility into a mechanical "buy low, sell high" behavior without requiring you to predict anything.

But only if you do it on purpose.

A good IPS states exactly how rebalancing happens:

- *Calendar rebalancing*: quarterly or annually.
- *Threshold rebalancing*: when allocations drift beyond a set band.
- Or a hybrid: check quarterly, act only if drift exceeds X%.

Notice what's missing: "rebalance when it feels like the right time." That sentence is just another way to say "rebalance when I'm comfortable," which is exactly when you least need it.

### 4) Define what you are allowed to do in a crisis

This is where the IPS becomes a behavioral firewall rather than a polite document.

Crisis rules should be blunt. Example:

> "I do not sell equities because of headlines, volatility, or forecasts. I will only reduce equity exposure if my time horizon changes, my need for liquidity changes, or my written plan changes."

That single sentence eliminates most panic selling. It doesn't eliminate fear. It eliminates the idea that fear is an instruction.

If you want a more structured guardrail, add a "cooling-off period" rule. A cooling-off period is simply a mandatory delay between impulse and action:

> "Any trade over $X requires a 48-hour waiting period and a written reason tied to my IPS."

The delay does something magical: it allows the emotional spike to pass. Markets can be volatile; emotions are often even more volatile.

### 5) Create a checklist for selling

Selling is not evil. Selling without criteria is.

A selling checklist prevents you from confusing discomfort with information. It can be as simple as a few gates you must clear:

- Has the original reason for owning it materially changed?
- Is this position now too large relative to the IPS cap?
- Do I need the money within the next 12 months?
- Am I selling to reduce risk, or to reduce anxiety?

That last question is the one people avoid because it's embarrassing. It shouldn't be. Anxiety is not a character flaw. It's a biological event. The point is to keep biological events from dictating financial structure.

### 6) Put guardrails where your hands actually are: automation

Automation is behavioral finance in its most practical form. If you automate the right behaviors, you don't have to summon discipline repeatedly. You summon it once, when you set the system up.

The most powerful automation is the boring one: automatic contributions. When money flows into the portfolio every paycheck, you are dollar-cost averaging—buying through good times and bad without having to "time" anything. You are also turning market downturns into a feature: the same contribution buys more shares when prices are lower.

If you want to feel what a system does, compare these two investors:

- Investor A contributes manually "when it feels like a good time."
- Investor B contributes automatically every payday.

Investor A is exposed to emotion every time they invest. Investor B is exposed once: when they choose the contribution rate. Over decades, that difference is enormous.

Another automation is automatic escalation: increasing contributions over time, ideally when your income rises. This is the quiet genius of programs like "Save More Tomorrow." It sidesteps the pain of cutting current consumption by linking higher savings to future pay increases. The investor doesn't have to feel deprived; they simply avoid upgrading their lifestyle as fast as their salary upgrades.

Lifestyle inflation is a compounding machine too—just pointed in the wrong direction.

Rebalancing can also be automated, depending on the account type and platform. Even when full automation isn't available, you can semi-automate by setting recurring calendar reminders and using pre-set rules. The key is that the rule exists before the emotion arrives.

### 7) Use pre-commitment: tie your own hands on purpose

Pre-commitment sounds dramatic, like you're planning to strap yourself to a mast. But you do pre-commitment all the time. You sign a lease. You schedule a dentist appointment. You set up a mortgage autopay. You are constantly limiting your future options because you know your future self will choose comfort.

Investing is one of the highest-return places to apply the same logic.

Pre-commitment tactics that work:

- *Account friction*: keep your "play money" account separate from your long-term portfolio, with different logins and no instant transfers.
- *Trading constraints*: require two-factor authentication, written memos, or a second person's review for large trades.
- *Withdrawal friction*: set up rules that any distribution requires a waiting period or a specific "allowed reasons" list.

These are not signs of weakness. They are signs you understand incentives. Casinos don't rely on customers to self-regulate. They design environments. You should too—except you're designing an environment that protects you from the casino in your pocket.

### 8) Make the plan observable: measure process, not mood

A plan you don't track becomes a plan you don't follow.

But "tracking" doesn't mean staring at daily P&L. That just recruits loss aversion and anxiety. Tracking should be process-based:

- Contribution rate (as a % of income).
- Current allocation versus target.
- Fees and taxes (are they controlled?).
- Rebalancing dates (did it happen?).

Process metrics are emotionally boring. That's their advantage. They make you feel less, which makes you do fewer dumb things.

### A short, usable IPS template

To make this concrete, here is a compact template you can adapt. It's written to be readable during a stressful market week.

```
INVESTMENT POLICY STATEMENT (1--2 pages)

1) Mission:
   - Primary goal:
   - Time horizon:
   - What this money is NOT for:

2) Risk and allocation:
   - Target allocation:
   - Allowed ranges:
   - Single-position caps (if any):

3) Contributions:
   - Automatic contribution amount/frequency:
   - Auto-increase rule (annual or tied to raises):

4) Rebalancing rule:
   - When I check:
   - When I act:

5) Trading and selling rules:
   - Cooling-off period for major trades:
   - Conditions under which I may reduce risk:
   - Conditions under which I will not sell:

6) Emergency rules:
   - Cash reserve target:
```

```
– If I feel panic, I will:
  (a) wait __ hours
  (b) reread this IPS
  (c) make no portfolio changes until then
```

If you can't fit your policy on two pages, you're probably using complexity as a form of comfort. Complexity feels like control. Control is comforting. But complexity also creates loopholes, and your future emotional self is a talented lawyer.

Simple rules are harder to argue with.

### The deeper point: stop negotiating with yourself

Most investing mistakes are not caused by a lack of intelligence. They're caused by a lack of *closure*. Every time you open your account and mentally renegotiate your strategy, you are re-running a high-stakes debate in an emotional courtroom. Markets provide fresh evidence every day. Your brain provides fresh feelings every minute. You will not win that case consistently for 30 years.

You need to close the case in advance.

That's what behavioral systems are: the closure you install while calm, so you don't have to invent it while terrified. They're not about being robotic. They're about being realistic about how humans behave under stress and designing around it.

### Actionable takeaway:

Within the next 60 minutes, write a one-page IPS draft and implement *one* piece of friction today. Not tomorrow. Today. Choose the friction that best matches your personal failure mode:

- If you panic-sell: add a 48-hour cooling-off rule for any equity sale and write it at the top of your IPS.

- If you chase trends: separate your "fun money" into a small capped account with no instant transfers.
- If you simply forget: automate contributions and set a recurring calendar reminder for rebalancing.

You don't need perfect discipline. You need a system that keeps you invested when your instincts would prefer relief. That's the behavioral firewall: not a heroic personality, but a well-designed environment.

Chapter 5

# Friction, Evidence, and Survival

*In the quest for wealth, we obsess over picking winners, yet the biggest fortunes are often built by those who simply refuse to lose. While markets offer returns, friction-in the form of fees, taxes, and bad behavior-relentlessly grinds them down. This chapter flips the script: instead of chasing alpha, we will hunt down the silent killers of compounding and learn the mathematics of survival. We'll explore why keeping your costs low is the only guaranteed return, how to read backtests without being fooled, and why avoiding total ruin is vastly more important than maximizing every percentage point of gain.*

## 5.1 The Mathematics of Fees and Drag

A fee is the most polite kind of wealth transfer. Nobody kicks down your door. No one sends a threatening letter. The money just leaves—quietly, automatically, and with the kind of regularity you normally associate with gravity. That's why fees are so dangerous: they don't feel like a decision you're making. They feel like the price of admission.

But in long-duration investing, *price of admission* is the wrong mental model. The better one is: a fee is a slice taken out of your compounding engine every single year, whether markets were kind to you or not. It doesn't merely reduce this year's return. It reduces next year's *starting line*. And that's where the math turns from annoying to brutal.

The easiest way to see it is to imagine two investors who are equally disciplined, equally diversified, equally patient. They buy the same market exposure. The only difference is the friction they agree to pay. Investor A captures the market's return. Investor B pays 2% per year in all-in recurring costs—expense ratios, advisory fees, platform fees, wrap fees, whatever name the industry chooses to print on the invoice. If the market delivers 8% per year gross, Investor A gets 8%. Investor B nets 6%.

That doesn't sound like a tragedy. Two percentage points is the kind of number people will spend more time negotiating on a mortgage rate they'll refinance in five years than on an investment plan they'll keep for fifty. Yet it's exactly the kind of small leak that sinks an otherwise seaworthy ship, because compounding is multiplicative. You don't lose 2% once. You lose 2% *of an ever-growing base*, and you lose the future growth on the money that leaked out.

The core formula is almost embarrassingly simple:

$$\text{Ending Wealth} = P(1 + r)^n$$

where $P$ is your starting principal, $r$ is the annual return, and $n$ is the number of years. If you want to understand fees, you don't need a new equation. You just need to accept that a fee lowers $r$. That's it. One tiny subtraction inside the exponent. And because it's inside the exponent, it's not tiny for long.

Put numbers on it. Start with $10,000. Let it run for 40 years.

- At 8%: $\$10{,}000 \times (1.08)^{40} \approx \$217{,}245$.
- At 6%: $\$10{,}000 \times (1.06)^{40} \approx \$102{,}857$.

Same market. Same patience. Same time horizon. One line ends north of $217k. The other barely clears $103k. The *fee* didn't cost you $\$10{,}000 \times 2\% \times 40 = \$8{,}000$. It cost you roughly $114k of ending wealth on a $10k starting stake. And that's before you even talk about contributions, which only increase the dollar value of the damage.

There's a particular cruelty to this arithmetic: the longer you succeed, the more you pay. A 2% drag on a $10,000 account is $200 in year one. A 2% drag on a $200,000 account is $4,000 in a single year. The better your habits, the more the system gets paid to watch you win.

Now here's the visual version of that story—two lines with the same starting point, slowly separating, then ripping apart so widely that the gap begins to look less like a difference in fees and more like a difference in lives.

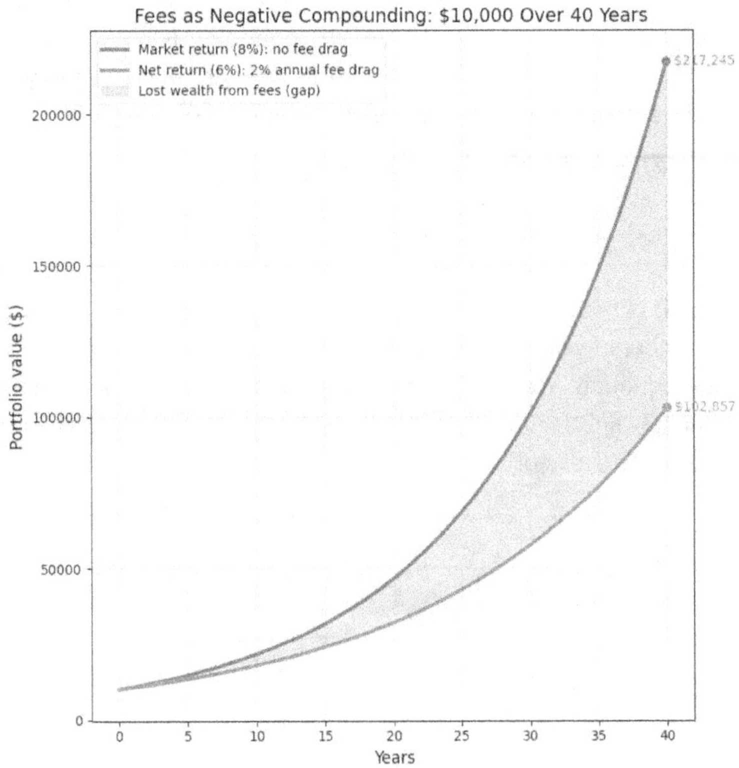

If you want a faster gut-check, you can abuse the Rule of 72. Normally it estimates doubling time: $72/r$ (with $r$ in percent). At 8%, money doubles about every 9 years. At 6%, it doubles about every 12 years. That three-year delay doesn't sound like much until you realize what it does over four decades: 40 years contains roughly 4.4 doublings at 9-year pace, but only 3.3 doublings at 12-year pace. One investor gets an extra *whole doubling* and change. That's not a rounding error; that's the difference between "comfortable" and "comfortable with options."

The deeper point is that fees don't behave like a one-time haircut. They behave like a permanent reduction in the engine's horsepower. You're not shaving the end result; you're lowering the growth rate that generates the end result. Once you see it that way, the most common fee conversations start to sound a little strange.

Consider the phrase, "It's only 1%." Only 1% is the kind of thing you'd say about a sales tax. It's not the kind of thing you should say about the exponent in the most powerful equation in personal finance.

Or consider how performance is often marketed: *gross* returns with a footnote. The footnote matters because your retirement will be funded with net returns, not footnotes. A manager who earns 8% gross in an 8% market has delivered you the market *minus* the bill. You took the market risk and paid extra for the privilege.

This is where costs become more than a personal-finance hygiene point. They become a structural advantage. A low-fee investor doesn't need to be smarter than the market. They don't need to find the next superstar manager. They don't need to time anything. They just need to *not leak*. In a game where most participants leak, not leaking is an edge.

The logic is unromantic but incredibly powerful: before costs, the average dollar invested must earn the market's average return. That's not philosophy; it's arithmetic. After costs, the average investor must underperform the market by roughly the average cost paid to access it. In aggregate, fees and trading costs don't disappear into the ether. They get paid to someone. And they come out of investors' returns.

That's why fee minimization is one of the only levers in investing with a near-guaranteed positive expected value. Markets may give you 8%, or they may give you 2%, or they may give you -30% in a year that tests your character. But if you negotiate your fee down from 1.5% to 0.15%, you didn't "predict" anything. You didn't "outsmart" anyone. You simply chose a better deal for the exact same economic exposure. It's like paying less interest for the same mortgage or paying fewer calories for the same meal: the savings are real regardless of how the world behaves.

There's another subtlety worth making vivid: fees are asymmetric. If markets return 8% and you pay 2%, you keep 6%. If markets return 2% and you pay 2%, you keep 0%. If markets return 0% and you pay 2%, you keep -2%. The fee doesn't care what kind of year you had. You could call that "reliability." The more honest name is "relentlessness."

This relentlessness also makes high-fee strategies psychologically tricky, because they can feel fine *for a while*. In the early years, the dollar amounts are small. A $200 annual drag is easy to ignore when your attention is on contribution habits and account setup. The pain is delayed. Like many delayed consequences, it's conveniently ignored until it is no longer convenient.

By the time the account is large enough for the fee to be obviously expensive, you may be socially and emotionally locked in. The advisor is "my person." The fund is "what I've always owned." The statement is "how I track my progress." Fees exploit the same human weakness that credit cards do: we adapt to the monthly number and fail to visualize the lifetime cost.

So it helps to translate percentages into a language your brain respects: *years of your life*. A recurring fee effectively shifts your financial clock backward. Two investors can both start at 25, invest for 40 years, and retire at 65. The low-fee investor's portfolio might look like a 65-year-old's money. The high-fee investor's portfolio might look like a 60-year-old's money. That's not a metaphor; that's the compounding curve. The fee steals time, and time is the only ingredient you can't earn more of.

To make the idea even sharper, imagine you were offered a deal:

> Pay me 2% of your portfolio every year, and I'll *maybe* beat the market.

You'd be right to ask: "Beat it by how much, and how often, and after taxes, and for how long?" Because the hurdle isn't 2%. The hurdle is 2% *compounded*. The manager has to overcome not just the annual subtraction, but the fact that the subtraction shrinks the base that future returns are earned on.

That's why the most dangerous comparison in investing is a simple one-year performance chart. Over a year, 2% of drag can be masked by noise, luck, and whatever factor happened to be in fashion. Over decades, 2% is a gravity field.

If you want a practical way to evaluate any recurring cost—expense ratio, advisory fee, platform fee—convert it into an "effective wealth tax" on your future. A quick-and-dirty approximation for the fraction of terminal wealth you keep is:

$$\left(\frac{1+r-\text{fee}}{1+r}\right)^{n}$$

Using $r = 8\%$, fee $= 2\%$, $n = 40$:

$$\left(\frac{1.06}{1.08}\right)^{40} \approx 0.47$$

Meaning you keep only about 47% of what you would have had. Put differently: a 2% annual fee can plausibly hand away *more than half* of your potential ending wealth across an investing lifetime. That's not fear-mongering. That's what exponentials do when you nudge the growth rate.

Fees also interact with behavior in a way that rarely makes the brochure. The more friction you carry, the more you need everything else to go right. You need better-than-average manager selection. You need to avoid switching at the wrong time. You need to stay invested through underperformance. You need to not succumb to the very human urge to fire a strategy after it disappoints you and hire it back after it redeems itself—the classic buy-high, sell-low loop. In other words, high fees don't merely tax your returns; they narrow your margin for error as a human being.

There's a reason this matters for survival. Compounding isn't fragile because markets are always evil; it's fragile because investors are. Anything that reduces your expected return also increases the probability that you abandon the plan during a long, discouraging stretch. A lower-fee portfolio doesn't just have better math; it has better *psychological runway*. It can underperform your hopes without underperforming your needs.

None of this implies you should never pay for advice or specialized management. Paying for expertise can be rational when it changes outcomes you actually control: tax planning, estate structure, insurance analysis, behavior coaching, withdrawal strategy design, and keeping you from detonating your plan during a crisis. The acid test is whether the fee buys a service that can plausibly be worth the drag *and* whether it's priced in a way that doesn't automatically scale up as you succeed. If your financial life is getting simpler while your account grows and the bill rises in perfect proportion, you're not buying complexity management. You're funding someone else's compounding machine.

A final practical wrinkle: the number you see labeled "expense ratio" is often not the only friction in the system. Trading costs, bid-ask spreads, cash drag, taxes triggered by turnover, and layers of fees in fund-of-funds structures can all add drag without a neat, single-line disclosure. The investor's job isn't to become paranoid; it's to become numerate. Whenever possible, ask for the all-in, recurring number and then run the 40-year math on it. If someone can't or won't help you estimate the true drag, that's information.

Actionable takeaway: pick one investment you own (a mutual fund, ETF, managed account, or advisory relationship) and write down its all-in annual drag as a single percentage. Then run this sentence with your real numbers: "Over 40 years, if markets return 8%, this drag turns my return into ___%, and $10,000 becomes about $___ instead of $___." If the difference makes your stomach drop, good—your intuition just caught up with the exponent.

## 5.2 Tax-Aware Compounding

Most investors can tell you their fund's expense ratio to two decimal places. Fewer can tell you what percentage of their lifetime gains will end up funding their state, the IRS, and the occasional surprise surcharge that appears when your income gets "too successful." That imbalance is understandable: fees show up as a line item. Taxes show up as a mood.

The trouble is that, over a multi-decade investing life, taxes are usually the biggest recurring drag on wealth that *looks* like it's compounding just fine. You watch the account rise, you feel responsible, you keep contributing. Meanwhile, a second invisible account is also rising: the portion of your returns that are quietly being pre-claimed. You don't see it on your brokerage dashboard because it isn't a separate bucket. It's simply the share of your growth you won't get to keep.

The emotional mistake is treating taxes as weather: inevitable, uncontrollable, something you complain about and then move on. The mathematical mistake is treating taxes as a one-time toll at the end. For many portfolios, the real damage comes from *annual taxation*—small bites taken each year that reduce the base available to compound next year. Just like fees, taxes can be negative compounding.

Unlike fees, taxes have loopholes, deferrals, and account structures that let you decide *when* the bite happens and *what* gets bitten.

Here's the frame that changes how you plan: the market gives you a gross return; your real life runs on an after-tax return. Gross returns are bragging rights. After-tax returns are groceries, tuition, freedom, and the ability to stop doing work you no longer want to do.

*A simple thought experiment clarifies the whole game.* Imagine you own an asset that grows 8% per year for 30 years. In World A, you pay tax every year on the growth. In World B, you pay the tax once at the end. Even if the tax rate were identical, World B tends to win because it lets the entire pre-tax amount keep compounding. Deferral isn't a cute perk; it's a structural advantage. It's like the difference between planting a tree in soil versus planting it in soil while someone trims 20% of the new growth every year.

That's why "tax-aware compounding" isn't just about tax brackets. It's about *friction timing*. If fees are the drag you can see on the prospectus, taxes are the drag you can often control with architecture.

### The three buckets you actually live in

In practice, most U.S. households end up with money in some mix of three account types:

- *Taxable accounts*: regular brokerage accounts. Dividends and interest can be taxed along the way, and realized capital gains are taxed when you sell.
- *Tax-deferred accounts*: traditional 401(k)s and traditional IRAs. You generally get a deduction (or pre-tax contributions), the account grows without annual taxation, and then withdrawals are generally taxed as ordinary income when you take distributions.[1]

[1] If you made nondeductible traditional IRA contributions, part of the withdrawal may be nontaxable basis tracked via Form 8606.

- *Tax-free accounts*: Roth IRAs and Roth 401(k)s (subject to rules). You pay tax up front, growth is sheltered, and qualified withdrawals are generally tax-free.

If those labels feel abstract, translate them into a more human question: *When does the bill arrive?* Taxable accounts send smaller bills repeatedly. Tax-deferred accounts postpone the bill. Roth-style accounts prepay the bill and (if rules are met) stop sending bills altogether.

The investing industry loves to talk about returns. Taxes force you to talk about *ownership*. What portion of your future balance is actually yours, and what portion is a future obligation with your name on it? In a traditional 401(k), you might feel like you have $500,000. In reality, you have $500,000 *minus whatever your future tax rate will be on withdrawals*. The money is real, but it's not all yours in the way a taxable account's cost basis is yours.

This doesn't mean tax-deferred accounts are bad. Quite the opposite: deferral is powerful. It just means you should stop treating account balances as comparable without translating them into after-tax terms.

### The compounding difference between ``taxed annually" and ``taxed later"

There's a quiet reason tax planning matters even for people who hate tax planning: annual taxation reduces your compounding rate in a way your brain can't feel month to month.

Suppose a bond fund yields 4.5%, and those distributions are taxed each year at ordinary income rates. In a taxable account, part of that return gets peeled off and sent away, leaving a smaller net yield to reinvest. Contrast that with holding the same bond exposure in a tax-deferred account: the full yield keeps reinvesting. The difference can look small in year one and grotesque over 25 years, especially as your account grows and the dollar size of the tax bite increases.

Equities can be friendlier in taxable accounts *when* they're implemented in a tax-efficient way: broad-market index funds and many ETFs tend to have lower turnover (fewer taxable events created inside the fund) and can distribute a meaningful portion of their return as unrealized appreciation rather than taxable income. That's not magic; it's simply the difference between growth that forces a tax event now and growth you can choose to realize later.

The result is a practical investing truth that feels backwards the first time you hear it: the same asset class can be a great long-term holding, but a terrible long-term holding *in the wrong account*.

### Asset location: not what you own, but where you put it

*Asset location* is the craft of placing different types of investments in the accounts that treat them most kindly. Asset allocation answers "How much stock and how much bond do I want?" Asset location answers "Where should the stocks and bonds live?"

This is one of the few areas in investing where the edge doesn't come from predicting markets. It comes from understanding the rules of the arena.

A simple heuristic captures most of the benefit:

- Put *tax-inefficient* assets in tax-advantaged accounts.
- Put *tax-efficient* assets in taxable accounts.

"Tax-inefficient" usually means "throws off income that gets taxed at higher rates or triggers frequent taxable events." Think taxable bonds, bond funds, REITs, and high-turnover active strategies. "Tax-efficient" often means "returns show up more as price appreciation and qualified dividends." Think broad equity index funds and many tax-efficient ETFs.

The hidden elegance is that asset location can add return without taking more market risk. It's a higher after-tax expected return created by less leakage, not by a spicier portfolio.

Of course, real life is messier. You may only have one type of account. Your employer plan may have limited options. You may be early in your career (lower bracket) or late (higher bracket). You may have state taxes. You may have charitable giving plans. The point isn't to chase a perfect location map. The point is to stop pretending location doesn't matter.

### Dividends: ``qualified" versus ``ordinary" is not a technicality

Dividends feel like free money because they arrive as cash. Taxes remind you they're simply one form of return.

In the U.S., some dividends can be *qualified* dividends, which may be taxed at the preferential long-term capital gains rates (often 0%, 15%, or 20% federally, depending on taxable income). Other dividends are *ordinary* (non-qualified) and are taxed like wages. That difference is not cosmetic. It changes the after-tax yield you actually get to reinvest.

What determines whether a dividend is qualified? There are rules around the type of company and how long you held the shares around the dividend date. The key behavioral lesson is simpler: if you trade frequently, you can accidentally convert what could have been preferentially taxed return into regularly taxed return. You don't need to understand every line of the tax code to internalize the investing takeaway: *high turnover often means high tax drag.*

A portfolio that produces the same pre-tax return can produce a very different after-tax outcome depending on whether it distributes mostly qualified dividends, ordinary income, or short-term capital gains. The taxes don't care that you "made money." They care *how* you made it.

### The NIIT: the surcharge that makes ``my bracket" an incomplete answer

High earners sometimes plan taxes as if the only question is, "Am I in the 24% bracket or the 32% bracket?" The U.S. tax system has a habit of adding extra layers once your income crosses certain thresholds.

One of those layers is the Net Investment Income Tax (NIIT), an additional 3.8% tax that can apply to investment income above certain modified adjusted gross income thresholds (commonly cited thresholds are $200,000 for single filers and $250,000 for married filing jointly). In other words, once your income is high enough, some of your investment income may face an extra toll.

You don't need to memorize the details to learn the investing lesson: tax drag is not linear. It can jump. And when it jumps, the value of tax-efficient implementation and asset location tends to rise with it.

### Roth versus traditional: the tax-rate bet you're always making

Roth and traditional accounts aren't just different containers. They represent a wager about your lifetime tax profile.

A traditional contribution is typically valuable when your current marginal tax rate is high relative to what you expect in retirement. You get a deduction now (or pre-tax contributions), you allow compounding to occur without annual tax, and you pay later—often after you've stopped earning wages. A Roth contribution is typically valuable when your current marginal tax rate is low relative to what you expect later, or when you value the certainty of tax-free growth and withdrawals.

The reason this matters for compounding is that taxes are an ownership claim. Paying 25% tax today on $10,000 and then compounding the remaining $7,500 tax-free can beat compounding the full $10,000 and then handing over a larger percentage at the end—or it can lose. The answer depends on future rates, withdrawal patterns, and policy risk. That uncertainty makes people freeze.

Freezing is usually worse than choosing an imperfect mix.

A surprisingly robust approach is simply to build *tax diversification*: having some money in taxable, some in tax-deferred, and some in Roth-style accounts (if available). It's the tax version of not putting all your eggs in one policy basket. With multiple buckets, you can often manage your taxable income in retirement by choosing which bucket to draw from, potentially controlling which brackets you fill and which surtaxes you trigger. The math is less about beating the IRS and more about buying flexibility in the decade when flexibility is priceless.

### Capital gains: the most underappreciated option in a taxable account

A taxable account has one superpower that doesn't get enough respect: you can choose when to realize capital gains.

If you hold a broadly diversified, low-turnover equity fund for decades, most of the return can arrive as price appreciation. As long as you don't sell, you generally don't realize the gain. That deferral can mimic some of the compounding benefits of tax-advantaged accounts, especially if your dividends are modest and largely qualified.

That's why many long-term investors treat taxable accounts not as a consolation prize but as a strategic bucket: a place for tax-efficient equity exposure, harvesting losses in bad years to offset gains, donating appreciated shares to charity when it fits, and controlling the timing of sales. It's not as clean as a Roth. It's also not as bleak as people assume when they only look at the word "taxable."

The psychological shift is to see tax planning as part of portfolio design, not as paperwork you do after the investing is over. When you decide what to buy and where to hold it, you're deciding how much of your return is allowed to compound uninterrupted.

### A practical way to think about ``tax drag" without becoming a tax accountant

The phrase *tax drag* means the amount by which taxes reduce your return over time. Investors often treat it like a nuisance because it's hard to see. A simple way to make it concrete is to convert it into basis points of return the way you did with fees.

Ask a pointed question: *If two funds hold similar exposures, how much return do I lose to taxes each year because of distributions and turnover?* Sometimes you can infer it by looking at distribution yields, turnover, and your marginal rates. Sometimes tools and tax-cost ratios can help. You don't need perfect precision. You need an honest estimate.

Then do the same mental move you made with fees: put it in the exponent. A portfolio that returns 7.5% after tax instead of 8.0% after tax isn't "only" 0.5% worse. Over 30 years, it can be the difference between reaching a goal and renegotiating your life around missing it.

Taxes are especially painful because they tend to attack the very things disciplined investors do well: earning more, saving more, and watching the account grow. When the account is small, taxes are small. When the account becomes meaningful, taxes become strategic.

### The ``right" answer is personal, but the direction is universal

The fine details depend on your bracket, your state, your time horizon, your planned giving, your career stability, and whether your employer offers a match. But the direction of travel is universal: reduce recurring tax leakage, maximize deferral where it's advantageous, and place assets in the accounts that tax them most gently.

The easiest mistake is focusing on the dramatic decisions ("Should I buy stocks or bonds?") and ignoring the quiet ones ("Which account should hold my bonds?"). The dramatic decisions get headlines. The quiet ones get outcomes.

Actionable takeaway: make a one-page map of your accounts and label each holding as *tax-efficient* or *tax-inefficient* in plain language ("low-turnover stock index," "bond fund that throws off interest," "REIT fund," "active fund with high turnover"). Then do a single swap on paper: move your most tax-inefficient holding into your most sheltered account (traditional or Roth), and move your most tax-efficient holding into taxable. If you can't execute the swap because of limited options or embedded gains, you've still learned something important: your constraints aren't abstract—they're where tax drag lives.

## 5.3 Implementation Vehicles: Funds and ETFs

A portfolio is an idea until you try to buy it.

On paper, "own the global stock market and a high-quality bond allocation" looks clean and almost moral—a vow of humility in a world of prediction. In reality, you implement that vow through products with plumbing, policies, and little toll booths hidden in the corners. The market return is the meal; the vehicle is the plate it's served on. If the plate is cracked, you'll spend decades wondering why your dinner keeps leaking onto the table.

The most common cracked plates come in two shapes: mutual funds and ETFs (exchange-traded funds). Both can give you diversified, low-cost exposure. Both can also be packaged in ways that quietly reintroduce the frictions you worked so hard to avoid: recurring fees, trading costs, tax drag, and behavioral traps.

The goal isn't to pick a team. It's to understand the physics of each vehicle well enough that you stop paying for problems you didn't know you bought.

## Two wrappers, same underlying engine

At the risk of sounding too simple, both mutual funds and ETFs are wrappers around a portfolio of securities. The wrapper determines *how you transact*, *how costs show up*, and *how taxes tend to behave*. The underlying engine—a basket of stocks or bonds—is what produces the return.

That wrapper distinction matters because friction often arrives through the wrapper rather than the holdings. Two products can hold essentially the same stocks and still deliver different investor experiences, especially around purchase mechanics and tax timing.

Mutual funds are typically bought and sold once per day at the fund's net asset value (NAV), the value of the underlying holdings divided by shares outstanding, calculated after the market closes. ETFs trade all day on an exchange like a stock. That single sentence drives most of the real-world tradeoffs.

If you've never cared about execution mechanics, you're in good company. Most long-term investors shouldn't care in the way day traders care. But you should care in the way a pilot cares about fuel lines: not obsessively, just enough to avoid a preventable failure.

## The visible price tag: expense ratios are real, but not the whole bill

The expense ratio is the annual operating fee charged by a fund, expressed as a percentage of assets. It's the cost that shows up in the brochure, the thing everyone quotes, the fee that feels like it should settle the matter.

It doesn't.

An expense ratio is like the sticker price on a car. It tells you something important. It doesn't tell you about financing, maintenance, or whether the dealer added a "documentation fee" that somehow costs more than a small refrigerator.

With funds and ETFs, the sticker price is the expense ratio. The maintenance and doc fees show up as trading frictions, cash drag, and taxes. Sometimes they're small. Sometimes they dominate.

So keep the expense ratio obsession, but upgrade it: treat it as *necessary but insufficient*. Low fee is table stakes for a compounding machine, not the finish line.

Mutual funds add a special twist: share classes. The same underlying fund can have multiple versions with different expense ratios and different sales charges (loads). Investors can end up in the expensive share class not because they chose higher costs, but because someone else chose distribution compensation. The portfolio may be identical. The investor outcome is not.

A practical habit: whenever you consider a mutual fund, look for other share classes of the same fund name. If one is dramatically cheaper, ask why you're not being offered it. "Because that's what we use" is not an answer. It's a confession.

### The trading price tag: spreads and premiums are fees with a different accent

ETF advocates will tell you they're cheap. Mutual fund advocates will tell you mutual funds are simpler. Both are right often enough to be dangerous.

ETFs introduce a friction that mutual fund investors rarely see: the bid-ask spread. The *bid* is the price someone is willing to pay. The *ask* is the price someone is willing to sell for. The difference is the spread—and when you trade, you often pay it implicitly.

In calm, liquid markets for large, heavily traded ETFs, spreads can be tiny. In niche ETFs, stressed markets, or less liquid holdings, spreads can widen. That widening is a cost you feel the moment you buy and sell. It's not a line item labeled "fee." It's a worse price.

ETFs can also trade at a premium or discount to NAV—meaning the ETF share price can drift above or below the value of its underlying holdings. This sounds like a design flaw until you understand the mechanism that usually pulls it back: the creation/redemption process run by authorized participants (APs), large institutional players that can exchange ETF shares for the underlying basket of securities (or vice versa). When the ETF price deviates, APs have an incentive to arbitrage the gap, which tends to keep the ETF price tethered to NAV.

Usually. Not always. And "usually" is the word that matters in survival-focused investing.

If you're a long-term investor making periodic contributions, you don't need to become an AP expert. You do need a simple instinct: if an ETF owns something that is hard to trade (thinly traded bonds, complex derivatives, niche international exposures), expect the ETF's trading frictions to show up at the worst possible moments—the very moments when your behavioral system is already stressed.

Mutual funds avoid intraday spreads because you transact at NAV. But they can have their own transaction frictions, including redemption fees in some cases, and the operational reality that your purchase executes after the close. Convenience is a real feature when you're trying to automate discipline.

### The market mechanics you don't see: ``cash drag" and the silent dilution problem

A fund that promises to hold a market exposure still has to handle the messy human fact that investors add and withdraw money at inconvenient times.

Mutual funds, in particular, may need to keep some cash on hand to meet redemptions, which can create *cash drag*: a portion of the portfolio sits in cash (or cash-like instruments) rather than in the intended assets, which can slightly lower returns in rising markets. Cash drag isn't always large, but it's a perfect example of implementation reality: your theoretical portfolio has no operational needs; your real portfolio does.

Mutual funds can also face a less intuitive issue: when many investors sell, the fund may have to sell securities to raise cash. Those sales can realize capital gains inside the fund. In a taxable account, those gains may be distributed to remaining shareholders, who then owe taxes—even if they personally didn't sell anything. This can feel like being handed someone else's bill.

ETFs, through their creation/redemption mechanism, can often manage flows differently. When redemptions happen in-kind (securities are handed out rather than sold for cash), the fund may be able to reduce the need to realize gains inside the portfolio. This structural difference is one reason ETFs are often described as tax-efficient.

That said, the tax advantage is not a law of nature. It's an artifact of rules and mechanics. Tax rules can change. Market structure can stress. Products can be engineered in ways that turn the clean ETF narrative into a more complicated story. A long-duration investor should love efficiency and remain suspicious of permanence.

## The ``ETF tax efficiency" story: powerful, but not a free lunch

ETFs are widely associated with tax efficiency because many can redeem shares in-kind, potentially limiting capital gains distributions. In plain English: instead of selling appreciated securities and distributing taxable gains, the ETF can hand out securities to an institutional redeemer, pushing the tax consequence outward. The investor often experiences more of the return as deferred capital appreciation rather than annual taxable distributions.

For investors holding in taxable accounts, this can be meaningful. It's not a mere optimization. It can change the compounding path by reducing the frequency of tax payments, which preserves the compounding base.

But don't let "tax-efficient" mutate into "tax-proof." Some ETFs distribute capital gains. Bond ETFs distribute interest income that is typically taxed at ordinary rates. High-turnover or actively managed ETFs can generate more taxable events. Specialized ETFs can behave in ways that don't match the vanilla index ETF story.

There's also a broader policy debate: some scholars argue that ETF mechanics can enable more than deferral in certain cases, raising questions about whether future regulation could reduce the advantage. A compounding plan that requires a loophole to remain open for 40 years is not a plan; it's a hope with a suit on.

The pragmatic takeaway is boring and therefore useful: treat ETF tax efficiency as a *current* structural benefit that can improve after-tax compounding, especially for broad, low-turnover equity exposure in taxable accounts. Then build your plan so it still works if that advantage narrows.

### The behavioral trap: vehicles that invite tinkering

ETFs trade like stocks. That's both a feature and a psychological hazard.

There's a specific kind of investor pain that begins with, "It's nice that I can trade anytime." Intraday tradability is wonderful if it helps you avoid panic (you can place a limit order, you can control execution, you can spread purchases through the day). It's harmful if it turns investing into a reflex.

Mutual funds, with their once-a-day pricing and sometimes clunkier trading interface, accidentally provide friction that can protect you from yourself. Not always, but often. When you can't compulsively trade the moment you feel anxious, your best self has a chance to show up.

This is not an argument for inconvenience as virtue. It's an argument for respecting the way humans misbehave around buttons. A vehicle that lowers *financial* friction can increase *behavioral* friction, and vice versa.

So when choosing between a mutual fund and an ETF for the same exposure, don't only ask, "Which is cheaper?" Ask, "Which one makes it easier for me to stay boring?"

The cheapest portfolio in the world is expensive if it causes you to trade it into dust.

### Index mutual fund versus index ETF: the practical decision tree

When the underlying exposure is similar and costs are competitive, the choice often comes down to execution preferences and account type.

In a tax-advantaged account (401(k), IRA, Roth), the tax efficiency differences between ETFs and mutual funds matter less because the account is already sheltered. In that environment, simplicity and automation can dominate: a low-cost index mutual fund with automatic investment features can be a compounding machine because it reduces decision points.

In taxable accounts, implementation details matter more. Broad, low-turnover equity ETFs often shine because they combine low expense ratios with structural tax efficiency and intraday liquidity. But you still need to respect spreads and premiums/discounts, particularly for niche or illiquid exposures.

And then there's the third variable that people ignore until it bites: your broker's mechanics. Some platforms make mutual fund purchases awkward or fee-laden and make ETF purchases frictionless. Others do the opposite. A plan that's theoretically perfect but practically annoying is the kind of plan you stop maintaining when life gets busy.

### A quick ``hidden cost" checklist that actually fits in your head

Instead of memorizing every micro-detail, keep a short checklist you can run in five minutes:

- *Expense ratio*: Is it competitive for the exposure?
- *Turnover*: Does the strategy trade a lot (tax and trading implications)?
- *Bid-ask spread*: For ETFs, is it consistently tight?
- *Premium/discount history*: Does it stay close to NAV, especially during stress?
- *Distribution behavior*: Does it throw off taxable income or capital gains?
- *Automation*: Can you set contributions and rebalance with minimal willpower?

None of these require you to forecast returns. They require you to notice friction.

### A story investors recognize: the ``cheap fund" that was expensive anyway

An investor chooses a low-fee fund and feels proud—correctly. A year later, they notice the fund has distributed a large taxable capital gain. They're confused. "How can an index fund distribute gains?" or "Why did I get taxed when I didn't sell?"

Often the answer is implementation, not betrayal. The fund may not be as passive as it sounded. It may have had turnover due to index reconstitution, strategy shifts, or flows that forced trading. Or the investor may have bought a mutual fund structure in a taxable account where in-kind mechanisms aren't available the same way. The investor chose low fees and still got hit by a different form of drag.

This is how friction hides: you stomp out one leak and discover another. That's not a reason to give up. It's a reason to treat implementation as a first-class investing decision, not an afterthought.

The compounding machine isn't just your allocation. It's allocation plus vehicle plus account plus your own temperament.

Actionable takeaway: pick one holding you own (or plan to buy) and write a one-paragraph "implementation memo" for it: (1) what exposure it provides, (2) what you pay in expense ratio, (3) what the likely hidden costs are (spreads, turnover, distributions), and (4) why you chose mutual fund or ETF for *your* account type. If you can't explain it without hand-waving, you don't yet own the vehicle—the vehicle owns you.

## 5.4 Skepticism: Backtesting and Forecasting

The most seductive chart in finance is the one that looks like it can't lose. A smooth upward line. A tiny drawdown. A tidy explanation: "It works because human nature never changes," or "It exploits a structural inefficiency." The pitch doesn't feel like gambling; it feels like physics.

And that's the problem. Markets are not physics experiments. They're battlefields of incentives, fees, taxes, and crowded trades, with humans panicking in the middle. If you want your wealth to survive decades, you need a skill that sounds negative but is actually protective: skepticism. Not cynicism. Skepticism is curiosity with a seatbelt.

The practical reason to care is simple: any strategy can be made to look brilliant in hindsight. Your job is to figure out whether brilliance is likely to persist *after* costs, *after* taxes, *after* implementation friction, and *after* the strategy becomes popular enough to be noticed by every other smart person with a Bloomberg terminal.

## Survivorship bias: the cemetery you don't get to tour

Most investing data is like an award show: it features the winners and quietly edits out the nominees that vanished.

Survivorship bias is the error of judging a process by the survivors rather than the full population. In investing, it shows up when databases, articles, and marketing decks focus on funds and strategies that exist today—which automatically excludes the funds that were merged away, liquidated, or quietly shut down after disappointing performance. The survivors tend to look healthier than the original cohort because they *are* healthier. That's what survival means.

A famous World War II story makes this painfully intuitive. Engineers studied returning aircraft and mapped where bullet holes appeared. The obvious idea was to add armor where the holes were most common. Statistician Abraham Wald pointed out the real lesson: those planes returned *despite* being hit in those places. The missing data was the planes that didn't return—the ones hit in places that caused fatal damage. Armor should go where the survivors *weren't* hit.

That is exactly how fund performance marketing works when you aren't careful. You see the bullet holes on the survivors and draw the wrong conclusion.

### Survivorship Bias (the airplane problem)

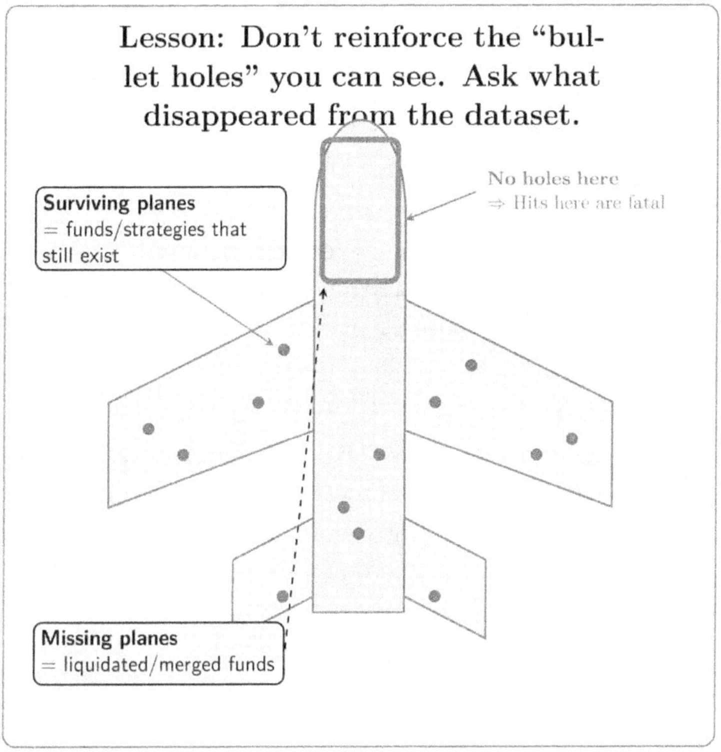


The investing translation is uncomfortable but liberating: when you read "Fund X has beaten its benchmark for 10 years," your first question shouldn't be "How do I buy it?" It should be "How many funds attempted something similar and are no longer here to be counted?"

If you only study the survivors, you will systematically overestimate skill and underestimate risk. It's not a moral failing; it's a data trap.

## Backtests: the market's most convincing fiction

A backtest is a simulation of how a strategy would have performed in the past. In principle, that's a sane thing to do. You wouldn't buy a car without looking at crash tests. You shouldn't buy a strategy without looking at history.

The trap is that markets offer only one realized history. Backtesting often feels like running experiments, but it's usually more like replaying a single movie and claiming you understand the entire genre.

The moment you backtest, you are tempted to improve the strategy. Adjust a parameter. Filter out a regime. Add a signal. Remove a year that "doesn't count" because the market was "irrational." Suddenly you're not testing; you're sculpting. You're shaping the strategy to fit the particular bumps and potholes of one historical path.

A backtest can be honest and still be misleading. If you try enough variations, you will find something that looks extraordinary. Not because the market is offering you a free lunch, but because randomness will eventually hand you a lucky streak that looks like a system.

That's data snooping: repeatedly mining the same dataset until chance starts to resemble skill. It's the statistical version of flipping coins until you get a long run of heads and then declaring yourself a master of physics.

The scary part is that the better you are at math, the more dangerous you can become. A sophisticated optimizer can produce a gorgeous curve that is, in effect, an artifact of selection. The curve is real. The *reason* you think it's real may be wrong.

### Overfitting: when the strategy learns your past instead of the market

Overfitting is what happens when a model captures noise rather than signal. In plain English: it learns the quirks of the sample period rather than the underlying structure of reality.

It's easiest to see with something non-financial. Imagine designing the "perfect" jacket size using measurements from 25 friends. If you tailor the jacket to fit every last shoulder asymmetry and posture quirk, you'll produce a jacket that fits your sample *beautifully* and fits a new person terribly. The jacket is overfit.

Strategies can be overfit in the same way. A rule set can end up encoding the accidental path of history: tech bubble, post-crisis rebound, low-rate decade, pandemic shock, liquidity wave. It can look brilliantly adaptive because it is, in a sense, customized to the past.

The deeper danger is psychological. Overfit strategies don't merely fail; they fail *confidently*. They tend to come with a story that explains why their historical success was inevitable. When they break, the story tends to mutate into: "The market changed." Sometimes the market did change. Often the strategy was never robust.

A practical sign of overfitting is parameter sensitivity: if a strategy's results collapse when you change a threshold from 12 to 11, or a lookback window from 200 days to 180, it may be perched on a razor blade. Real edges usually don't require the universe to be tuned to exactly your chosen integer.

### False precision: the spreadsheet's favorite lie

If you've ever seen a retirement forecast that projects your wealth to the nearest dollar in the year 2056, you've met false precision. The problem isn't the number. The problem is the implied confidence.

Forecasting in markets is like measuring fog with a ruler. You can do it. You can get a number. The number doesn't mean what it feels like it means.

Investment marketing often leverages this weakness by offering point estimates: "Expected return: 10.2%." "Maximum drawdown: 8.7%." "Probability of loss: 3.1%." These numbers can be computed; they can be formatted beautifully; they can also conceal the fact that the distribution is fat-tailed, regime-dependent, and subject to structural breaks.

Long-duration investors don't need prophetic precision. They need a range that is honest enough to build a plan that survives bad luck.

If someone gives you a forecast, ask two questions that quickly separate rigor from performance theater:

- What happens if your assumptions are wrong in the unfavorable direction for a decade?
- What happens after taxes, after fees, and after realistic trading frictions?

A strategy that only works in the center of the distribution is not a strategy. It's a fair-weather story.

## How to read a backtest without being hypnotized

Backtests are not useless. They're just easy to misuse. A good backtest is less like a victory lap and more like a stress test: an attempt to discover how the strategy behaves when conditions are ugly, boring, or adversarial.

Here are the kinds of questions that keep you honest without requiring a PhD:

**1) How many degrees of freedom were available?**

If the designer could choose among dozens of assets, multiple time windows, many filters, and several risk controls, the strategy had a large search space. Large search spaces produce accidental winners. You should demand stronger evidence.

**2) Was there out-of-sample testing?**

Out-of-sample means the strategy was designed on one period and then tested on a different period not used in design. This doesn't solve everything (markets are not stationary), but it's a basic hygiene check. If a strategy only "works" in the period used to build it, you haven't learned much.

**3) Were costs modeled like a grown-up?**

A strategy that trades frequently but assumes tiny friction is a fantasy. Spreads widen in stress. Liquidity disappears. Taxes show up at the worst times. A backtest that ignores realistic costs is not optimistic; it's incomplete.

**4) Is the edge plausible, not just statistical?**

A backtest should come with a mechanism story that survives cynicism. "It works because it worked" is not a mechanism. A plausible mechanism often involves risk compensation (you're taking a kind of risk that pays over time) or behavioral compensation (you're providing liquidity or absorbing discomfort that others avoid). If the story is "a secret pattern," assume the secret is that it's noise.

**5) Does the strategy require you to behave like a robot?**

Many strategies fail not because they stop working, but because humans can't hold them through inevitable underperformance. If the backtest includes three-year drawdowns and you know you'd bail after six months, the strategy's true expected return for you is not the backtest return. It's the return of whatever you'll switch into at the worst possible moment.

That last point matters because skepticism isn't just about math. It's about self-knowledge. A strategy that is fragile to your own emotions is not robust, even if it's robust to markets.

## The ``best strategy" problem: the winner's curse in research

There's a particular statistical cruelty in strategy selection: if you test many ideas, the one that looks best is likely to be the one that benefited most from luck. Even if every strategy has zero true edge, one will end up with a stunning backtest. Not because it's real, but because you asked the universe to produce a winner.

This is the winner's curse applied to research. The act of selecting the best-looking strategy inflates your estimate of its true performance. You don't just find the tallest person in a room and conclude they're tall; you find the tallest person in a city and conclude they're a giant.

The cure isn't despair. The cure is humility and process: fewer degrees of freedom, more insistence on out-of-sample evidence, more skepticism about smoothness, and more attention to implementation frictions that eat edges for breakfast.

## A skepticism toolkit that still leaves room for good ideas

Skepticism can become a personality, and that's not helpful. The goal is not to reject every strategy. The goal is to avoid being the raw material from which other people manufacture track records.

A practical middle ground looks like this:

If a strategy's pitch is "we found a pattern," treat it as guilty until proven robust. If the pitch is "we are deliberately taking a compensated risk" or "we are capturing a known premium in a low-cost way," you can evaluate it with a calmer brain. The second kind of claim is still uncertain, but it tends to be anchored in economics rather than curve-fitting.

If a strategy's evidence is a single backtest, treat it as a hypothesis. If the evidence includes multiple markets, multiple time periods, realistic costs, and a mechanism that makes sense, you may have something worth considering.

And then, even if it passes, ask the final survival question: *Is it worth the complexity relative to a cheap, broad, tax-aware implementation?* Many strategies are not competing against zero. They are competing against a boring, robust baseline that already compounds beautifully when you stop punching holes in it.

**Actionable takeaway:** the next time you see a strategy backtest (in an article, a fund pitch, or your own spreadsheet), write three sentences before you do anything else: (1) "What data might be missing from this story?" (survivorship bias), (2) "How many ways could the designer have tweaked this to look good?" (overfitting), and (3) "What's the after-fee, after-tax, after-friction return I'd actually experience?" If you can't answer those with reasonable confidence, the backtest is entertainment, not evidence.

## 5.5 Avoiding Ruin: Sizing and Fragility

The finance industry loves upside. Upside makes good slides, good podcasts, good careers. Survival is quieter. Survival doesn't sparkle; it *persists*. Yet for long-duration wealth, survival is the whole game, because compounding has an unforgiving rule: you can't compound from zero.

That sounds obvious until you notice how often people behave as if it isn't true. They take risks that are fine in a single year but fatal over a lifetime. They borrow against portfolios because "rates are low." They concentrate because "conviction is how you win." They treat volatility as an annoyance rather than a force that can permanently change the trajectory of an otherwise sound plan.

A useful mental shift is to stop thinking of risk as "How much can I make?" and start thinking of it as "How can I stay in the game long enough for my edge—or simply the market's long term tendency to rise—to matter?"

## The asymmetry that breaks people: losses don't heal at the same rate

Gains and losses look symmetric on a statement: +10%, -10%. Your nervous system treats them as wildly asymmetric (loss hurts more), and mathematics agrees with your nervous system.

If you lose 10%, you need an 11.1% gain to get back to even. If you lose 20%, you need 25%. If you lose 50%, you need 100%.

The deeper lesson isn't the arithmetic; it's the shape. Recovery requirements curve upward, then spike. As losses approach 100%, the required recovery approaches infinity because infinity is what it takes to resurrect a wiped-out account.

This is the math version of "don't die." And it's why risk management is not a conservative personality trait; it's a compounding prerequisite.

Here's the curve that should be burned into your investing instincts.

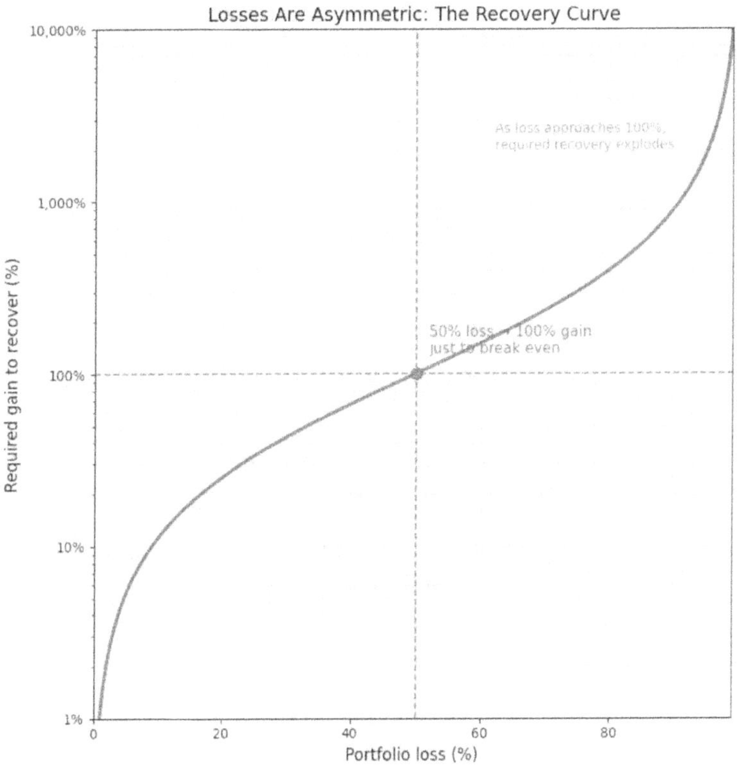


The picture is more than a curiosity. It explains a lot of real investor behavior that seems irrational until you accept that people fear large drawdowns because they should. A big drawdown doesn't just delay your goals; it can change what goals are even possible.

### Geometric drag: why volatility can lower long-run growth even when ``average returns" look fine

Here's another quiet asymmetry that punishes fragile portfolios: volatility reduces compound growth.

If returns bounce around, your long-term compound return (the *geometric* return) is typically lower than the simple average (the *arith-*

*metic* return). That gap is sometimes called *geometric drag*. It's not a fee you pay to someone else. It's a fee you pay to randomness.

A simple example makes it visceral. Imagine two years:

- Year 1: +50%
- Year 2: -50%

The average return is 0%. But $100 becomes $150, then becomes $75. You're down 25% even though the average looked neutral.

This is why the phrase "I can handle volatility" can be incomplete. Volatility isn't only an emotional problem. It's a mathematical one, because it interacts with leverage, concentration, and withdrawal timing in ways that can create permanent damage.

A portfolio that chases high average returns by accepting high volatility can easily end up with lower long-run compounding than a steadier portfolio. The irony is painful: you can take more risk and end up with less wealth.

### Ruin is not ``underperformance," it's game over

Most investors talk about risk as tracking error: "Will I underperform my benchmark?" That's a professional concern. A long-term investor's real risk is a different kind of error: a mistake so large it forces a permanent behavior change.

Ruin doesn't always mean literally hitting zero. In real life, ruin can mean:

- a drawdown that triggers a panic sale you never recover from,
- leverage that forces liquidation at the bottom,
- concentration that turns one bad outcome into a decade of clawing back,
- a retirement sequence that drains the portfolio early, making later recovery mathematically impossible.

The common feature is fragility: a system that can't tolerate a bad run.

Fragility is easy to miss because it often looks like strength in good times. Leverage makes you look smart until it makes you insolvent. Concentration makes you look decisive until it makes you dependent on one story staying true forever. Short-volatility strategies produce consistent gains until they produce a single catastrophic loss. The pattern is always the same: small steady wins, then a cliff.

Compounding hates cliffs.

## Sizing: the lever that separates bold from brittle

Position sizing is the unglamorous craft of deciding *how much* to allocate to any one risk. It is the bridge between "I have a good idea" and "I will still be solvent if I'm wrong."

Most investors think about what to buy. Professionals obsess about how big to buy it. The difference is survival.

Sizing matters because returns aren't linear in the presence of drawdowns. If you bet 10% of your portfolio on something and it goes to zero, you lose 10%. If you bet 50% and it goes to zero, you lose 50%, and the recovery math becomes punishing. If you bet 100%, you're done. The *same* idea can be a manageable mistake or a permanent loss depending on size.

This is why concentration both builds wealth and destroys it. Concentration is a tool. It is not a personality.

The right question isn't "Is this investment good?" It's "If this investment goes badly, what does that do to my life plan?" If the answer is "It breaks it," the size is wrong, regardless of how confident you feel.

### The Kelly temptation: maximizing growth versus maximizing survivability

The Kelly criterion is famous in gambling and has a seductive message in investing: there exists a bet size that maximizes long-run growth (technically, it maximizes expected log wealth). For a favorable game, Kelly tells you to bet more when you have an edge and less when you don't.

It sounds like the holy grail: math that tells you exactly how aggressive to be.

The catch is that full Kelly sizing can be brutally volatile. It can produce deep drawdowns even when the long-run outcome is favorable. In the real world, deep drawdowns are not merely uncomfortable; they can cause forced selling (margin calls), career risk (you fire your manager at the bottom), or self-betrayal (you abandon the plan).

So even if you believe in the logic of Kelly, practical investing often calls for "fractional Kelly": taking some portion of the theoretically optimal bet size to reduce drawdown risk. That's a fancy way of saying something very old: live to fight another day.

The meta-lesson is more valuable than the formula: the optimal strategy on paper is not optimal if you cannot hold it.

### Gambler's ruin: why ``positive expected value" can still kill you

There's a cruel misunderstanding that shows up in both casinos and markets: people think positive expected value means inevitable success.

But if your bet sizing is too aggressive, even a favorable game can ruin you before the edge has time to assert itself. That's gambler's ruin in spirit: once you hit the absorbing barrier (zero), the game ends. You don't get to average your way back.

Markets are not coin flips, but the survival logic transfers. If you take a portfolio that has a positive long-term expected return and you add leverage, concentrated exposures, or illiquidity, you can create paths where a bad sequence forces liquidation. The expected return doesn't matter if you can't stay invested.

This is why "I'm right in the long run" is not a risk management plan. The long run is a luxury granted only to those who avoid terminal outcomes in the short run.

### Fragility often enters through the side door: leverage, illiquidity, and ``safe" income

Fragility rarely announces itself as fragility. It arrives wearing the costume of efficiency.

Leverage is the obvious example, but not the only one. Any obligation that forces selling at an inconvenient time is a form of leverage. A portfolio that depends on an income stream that can be cut (high dividends from stressed companies, yield-chasing in credit) can be fragile if the investor treats that income as guaranteed. Illiquidity can be fragile when you *need* liquidity.

Taxes can add a subtler version of fragility. High-turnover strategies in taxable accounts can create recurring tax bills that must be paid regardless of whether the investor is reinvesting or withdrawing. In bad years, those bills can feel like insult layered on injury. A strategy that is fine in a tax-advantaged account can become psychologically and financially harder to hold in a taxable one.

The general rule is stark: if your plan requires everything to go reasonably well, it is not a plan. It is a prediction.

### Diversification: not a return maximizer, a ruin minimizer

Diversification is often sold as a way to improve risk-adjusted returns. That's true, but the more important point for a human investor is simpler: diversification reduces the chance that one narrative, one sector, one country, one management team, or one policy change ruins your decade.

Diversification doesn't prevent losses. It prevents *catastrophic* losses from concentrated exposures.

There's a psychological benefit too. A diversified portfolio gives you fewer reasons to stare at any single position like a verdict on your intelligence. It reduces the emotional amplitude of investing. Lower emotional amplitude tends to produce better behavior. Better behavior tends to produce better realized returns.

Your portfolio doesn't need to be perfect. It needs to be survivable.

## A concrete way to size risk: define your ``maximum tolerable drawdown" before you choose assets

Most investors discover their drawdown tolerance the worst way: through experience. The market gives them a drawdown, they learn their tolerance was lower than they thought, and they sell at the bottom to stop the bleeding. That is tuition you don't need to pay.

A more disciplined approach is to pick a number *in advance*: the maximum drawdown you believe you can tolerate without changing your plan. This number is not your ego. It's your behavior under stress.

Then build backward. If a portfolio is likely to experience 50% drawdowns, and you know a 30% drawdown would make you abandon the plan, the portfolio is too aggressive *for you*, regardless of how attractive its long-run expected return looks.

Sizing is not only about the size of one position. It's about the aggressiveness of the whole system.

Actionable takeaway: write a one-sentence "ruin rule" and make it operational. For example: *"No single position can cost me more than 10% of my portfolio, and no portfolio design is acceptable if a plausible drawdown would force me to sell."* Then test it: look at your largest positions (including any leveraged exposures, concentrated employer stock, or illiquid holdings) and calculate what happens if they fall 50% and 80%. If the recovery curve makes your stomach tighten, that's not fear. That's your compounding instinct finally speaking in math.

## 5.6 Stress Testing and Simulation

Averages are comforting. Averages are also how people get blindsided.

If you've ever met someone who did "everything right" and still had to delay retirement, you've seen the problem. They saved. They diversified. They kept fees low. They didn't panic-sell. And yet the calendar betrayed them: a brutal market showed up at the exact wrong time, early withdrawals carved permanent dents into the portfolio, and the compounding engine that looked unstoppable on a spreadsheet started coughing.

The villain wasn't ignorance. It was path dependence: in investing, the order of returns can matter as much as the returns themselves. When you're accumulating, bad early returns are annoying; they can even be helpful if you keep buying. When you're withdrawing, bad early returns can be lethal. Same average return. Different ending.

Stress testing is how you stop trusting the average and start respecting the path.

### Forecasting is a story; simulation is a stress test

Most financial plans start with a forecast: "Stocks return 8%, bonds return 4%, inflation is 2.5%, you retire at 65." Forecasts are not evil. They're a way to get a first draft.

The trouble is that forecasts tend to come as single lines. One future. One number for each input. One implied certainty. It's the same false precision problem wearing a friendlier outfit.

A simulation treats the plan less like a prediction and more like an engineering design. You don't ask, "What will happen?" You ask, "What could happen, and does my plan survive the ugly versions?"

That difference in posture changes your behavior. Forecasting invites confidence. Simulation invites humility. Humility keeps you solvent.

### Sequence-of-returns risk: the silent predator

Sequence-of-returns risk is the risk that a bad run happens early in a period when you're taking money out (or when you have a hard deadline). It's not about the long-term average return. It's about timing.

A simple example shows why this is not a niche worry for retirees only.

Imagine two investors who each earn the same average return over 10 years. Investor A gets the bad years first and the good years later. Investor B gets the good years first and the bad years later. If neither contributes nor withdraws, they end in the same place (because multiplication commutes). But if they're withdrawing—especially at a fixed dollar amount—the outcomes diverge sharply.

Withdrawals during a drawdown force you to sell more shares at depressed prices. That reduces future participation in the recovery. The portfolio doesn't just fall; it *shrinks its ability to rebound*. It's the drawdown-recovery curve in action, but now with a metronome tapping withdrawals on the way down.

Sequence risk is why "the market always recovers" can be true and still not save your plan. The market's recovery can arrive after your portfolio has already been weakened by forced selling.

### Monte Carlo simulation: not a crystal ball, a wind tunnel

Monte Carlo simulation sounds exotic. It's not. It's a method of generating many possible future paths by drawing returns from a model, applying your contributions or withdrawals, and seeing how often the plan succeeds.

Think of it like a wind tunnel for your financial life. Engineers don't test a new airplane design by flying it once on a calm day and declaring victory. They run it through conditions that include turbulence, crosswinds, and nasty edge cases, because airplanes fail at the edges. Portfolios do too.

A Monte Carlo simulation asks: if the future contains a mix of good years, bad years, and mediocre years in random order, what happens to me? Not just on average, but in the bottom 10%, the bottom 5%, the nightmares.

It's important to be honest about the limits. The model you choose matters. If you assume returns are normally distributed and independent, you may understate the frequency of extreme events and regime shifts. If you assume too much mean reversion, you may overstate recovery. If you ignore inflation spikes, you may overstate real spending power. Simulation is not truth. It's an organized way to be wrong in public.

That's still an upgrade from being wrong in private.

### What you should stress-test (because life will)

The best stress tests focus on the variables that actually break plans:

**Withdrawal rate and spending flexibility.** A plan that works only if you withdraw exactly 4.00% every year like a robot is a plan that hasn't met a human. Real life includes car replacements, medical expenses, helping family, and the occasional urge to do something joyful. A robust plan includes slack: the ability to reduce spending temporarily without feeling like failure.

**Inflation and real returns.** Nominal returns can look fine while real purchasing power erodes. Inflation doesn't just raise prices; it changes behavior. It makes people reach for yield, chase "inflation hedges," and take risks they don't understand. A stress test that ignores inflation is like a fire drill that assumes the smoke will be polite.

**Longevity.** The most optimistic failure mode is simply living longer than expected. That's a great problem to have and a brutal one to fund. A plan that works for 25 years can fail for 35, even with identical average returns, because compounding cuts both ways: withdrawals compound too.

**Concentration and correlated losses.** Diversification reduces the chance of catastrophic loss, but correlations rise in crises. Stress tests should assume some of your diversifiers will disappoint you at the same time. The goal isn't paranoia; it's to see whether you're quietly depending on a single regime.

**Implementation frictions under stress.** Bid-ask spreads widen. Premiums and discounts can behave strangely. Credit becomes tight. Taxes can be triggered by forced selling. A plan that looks fine in a frictionless spreadsheet can become fragile once you add the real-world mechanics of getting in and out.

When you add these stressors, something interesting happens: the conversation shifts from "What return do I need?" to "What risks am I implicitly taking, and are they risks I can afford?" That's the adult version of financial planning.

## Scenario analysis: the stories that models can't tell

Monte Carlo is great at exploring randomness. It's less good at capturing structural shifts and one-off shocks that are hard to model as random draws.

That's where scenario analysis shines. Scenario analysis is less about probability and more about plausibility. It asks: *What if something specific happens?*

Examples that are worth running even if you hate thinking about them:

- A 2008-style drawdown in the first two years of retirement.
- A decade of low real returns (not dramatic, just disappointing).
- Inflation spikes and stays elevated for several years.
- A sudden need for large cash outflows (health, family, job loss).
- A policy change that reduces a tax advantage you assumed would persist.

The point is not to guess the next crisis. The point is to discover whether your plan has a single point of failure.

Scenario analysis also has a psychological benefit: it makes the future feel less like a single looming unknown and more like a set of testable conditions. That reduces fear. Reduced fear improves behavior. Improved behavior improves realized returns.

### The most valuable output of a simulation is not a number

People often run Monte Carlo simulations to get a success probability: "My plan has an 87% chance of success." That number is seductive, and it can be useful as a rough gauge. But it's not the most valuable output.

The most valuable output is discovering *what makes the plan fail*. Failure analysis.

In a good simulation tool (or a well-built spreadsheet model), you can often examine the failure paths:

- Did failure happen because of a bad early sequence?
- Did it happen because inflation was higher than expected?
- Did it happen because withdrawals were too rigid?
- Did it happen because equity exposure was too low (inflation risk) or too high (drawdown risk)?

Once you know *why* failure happens, you can improve robustness in targeted ways: a slightly lower withdrawal rate, a cash buffer, dynamic spending rules, a shift in asset location, a reduction in concentrated exposures, a plan for part-time income, or a commitment to rebalance rather than flee.

That's the compounding mindset applied to planning: small adjustments that increase survival odds over decades.

### Model risk: the humility tax you should gladly pay

Every simulation bakes in assumptions. That's not a flaw; it's the nature of modeling. The flaw is forgetting that assumptions are there.

Treat your model like a map. A map is extremely useful. A map is also not the territory. If the map is wrong, you don't argue with the mountain. You update the map and adjust your route.

So include a humility tax in your planning:

- Assume your future returns might be lower than the historical average.
- Assume your future costs might be higher (healthcare has a talent for surprising people).
- Assume correlations behave badly in crisis.
- Assume you might not behave perfectly when things get scary.

You don't do this to be pessimistic. You do it to be resilient. Resilience is optimism that can survive contact with reality.

### A simple "robust plan" checklist that doesn't require fancy software

Even without running thousands of simulated paths, you can stress-test your plan with a few concrete questions:

- If the market drops 40% next year, what exactly do I do? (Not what do I *hope* I do.)

- If inflation runs hot for five years, which parts of my budget flex, and which don't?
- If I have to withdraw for two years during a downturn, what buffer exists?
- If my best-case return assumptions are wrong by 2% per year, does the plan still work?

Notice what this list does: it forces you to convert "risk" from a number into a set of decisions. Decisions are where wealth is made and lost.

Actionable takeaway: run one "bad sequence" drill in writing. Take your current portfolio and assume the next two years are -30% and -10% (in that order), followed by a normal recovery. Write down, step by step, how you would handle contributions, rebalancing, and any planned withdrawals. If you can't write a calm script you believe you'd follow, the plan isn't missing intelligence; it's missing robustness.

Chapter 6

# The Lifecycle of Money

*If investing is like climbing a mountain, accumulation is the ascent-grueling but straightforward. Retirement is the descent-technically easier, but far more dangerous. This chapter maps the changing physics of your money as it shifts from a growing snowball into a sustainable stream. We will move beyond simple savings rates to explore how the math of compounding changes when you stop adding capital and start relying on it, and why the strategy that got you rich is rarely the same one that keeps you wealthy.*

## 6.1 Accumulation: The psychology of Getting Money In

You inherit $120,000. Or your company gets acquired and your options finally become money. Or you've been "saving for a crash" and, for the first time in years, you're sitting on a cash pile that could actually change your life. That night you do the thing responsible adults do: you open a spreadsheet and try to be rational. And in the quiet glow of your monitor, a childish fear shows up anyway: *What if I invest it and the market drops tomorrow?*

That fear isn't ignorance. It's not even pessimism. It's the brain running a completely normal simulation: "I will feel stupid." Most investing advice tries to win the argument on math. Your brain is trying to win a different argument: minimizing future regret. If you don't acknowledge that, you'll do what millions of smart people do every year with a lump sum: you'll delay, "wait for clarity," and accidentally choose a third strategy nobody recommends—*permanent cash.*

There are only two honest ways to deploy a lump sum into volatile assets like stocks. You invest it now (lump-sum investing), or you spread it out over time (dollar-cost averaging, or DCA). DCA is simply a schedule: "I will buy $X per week/month for Y months until the cash is gone." Nothing mystical. Lump-sum investing is equally simple: "I will buy today." The debate gets heated because each side is optimizing for a different pain.

The math argument is straightforward: when the market has a positive expected return, earlier tends to beat later. If prices drift upward over time, delaying purchases means buying at higher average prices. That's why, in many historical comparisons, investing a lump sum immediately has outperformed DCA a majority of the time. It's the same reason leaving your money in a savings account "until things settle down" usually loses to investing: markets don't send invitations when they're done being messy.

But the psychological argument is just as straightforward: DCA is an emotional shock absorber. It turns one terrifying decision into twelve smaller ones. If the market drops right after you start, you don't feel like the person who "blew it." You feel like the person who is *executing the plan*. The same market outcome, completely different story in your head.

That story matters because accumulation is not a single decision. It's a habit built out of dozens of small, boring actions that you repeat when you're tired, busy, or scared. The hidden enemy of wealth isn't usually a lack of knowledge. It's *inertia*: money that could have been invested but wasn't, because your brain kept postponing discomfort. This is why the DCA vs. lump-sum question is really a question about whether you can reliably get money *in*.

*Regret minimization* is the cleanest lens for this choice. Instead of asking, "Which strategy wins more often?" ask, "Which strategy is least likely to make me abandon my plan?" A plan that is slightly suboptimal but actually executed beats a perfect plan that stays in a notebook.

### The two kinds of regret you're buying (whether you admit it or not)

When you invest a lump sum, you're exposed to *timing regret*. If the market falls next week, the pain is immediate and vivid. Your brain doesn't calculate long-term expected returns; it replays the moment you clicked *Buy*. That's not a minor annoyance—it can poison your willingness to invest future dollars. One bad lump-sum experience can make someone "a cautious investor" for a decade, which is a polite way of saying they'll underfund their own future.

When you DCA, you're exposed to a different regret: *opportunity regret*. If markets rise during your DCA period, you'll watch your uninvested cash miss gains in slow motion. This regret is quieter and, for many people, easier to tolerate because it doesn't come with a single shameful timestamp. There's no cinematic moment of failure. It's more like a gentle tax you pay for emotional comfort.

Here's the twist that makes this psychologically slippery: timing regret feels like *your fault*. Opportunity regret feels like *bad luck*. Humans are exquisitely sensitive to blame, especially self-blame. So even if the average dollar difference favors lump-sum investing, many investors will still choose DCA because they are trying to avoid the specific flavor of pain that makes them do something destructive later.

And destructive later is the real threat. In accumulation, mistakes compound, but not only through returns. Behavior compounds, too. Your future contributions are a stream of decisions, and your emotional state today influences whether you keep showing up.

### The market isn't only risky; your identity is on the line

Most people think they fear volatility. They don't. They fear the identity wound that volatility can cause.

If you lump-sum invest and the market drops, your brain doesn't merely register a number going down. It offers a narrative: *I am the kind of person who buys at the top. I knew I wasn't good at this.* That story is sticky. It can turn investing into a referendum on your intelligence.

DCA changes the narrative architecture. A drop becomes evidence that your next scheduled purchase will be at a lower price. Suddenly the same volatility can be interpreted as helpful: "Great, I'm buying more shares cheaper." This reframing isn't a trick; it's accurate for someone still accumulating. The market falling early in a DCA schedule can be beneficial because it means more of your purchases occur at lower prices. But even when the math is a wash, the story protection is real.

A useful way to think about DCA is as a *regret hedge*. You are giving up some expected return (in many environments) to reduce the chance of an emotionally catastrophic outcome that changes your behavior.

That might sound soft, but it's brutally practical. If a strategy reduces the odds you'll freeze, abandon your plan, or swear off stocks forever, it can be the higher-return strategy *for you* over a lifetime.

### Lump sum vs. DCA: the real comparison is ``time in the market" vs. ``time on your side"

Lump-sum investing maximizes *time in the market*. That phrase is popular because it's true: with positive expected returns, having your dollars exposed sooner is an advantage.

DCA maximizes *time on your side*. Not your dollars' time—your psychological time. DCA gives you days and weeks to acclimate. The first purchase doesn't have to carry the full emotional weight of the entire sum. You get to practice being an investor while the stakes are still scaling up.

This is why the DCA debate can feel strangely moral online. One side sounds "disciplined" and "rational." The other sounds "cautious" and "gentle." But neither is a virtue contest. It's a design problem: what deployment method makes it most likely that the money actually gets invested and stays invested?

There's also a subtle third element: DCA can protect you from *your own desire to be clever*. Many investors say they're doing DCA, but what they're actually doing is *disguised market timing*:

> "I'll invest $10,000 per month, unless the market looks scary, in which case I'll pause."

That isn't DCA. That's a flexible excuse generator. Real DCA is boring and automated. The whole point is to stop negotiating with yourself.

### A concrete way to choose: set your ``panic threshold" before you invest

Most investors make the DCA vs. lump-sum decision by asking, "What do I think the market will do next?" That's a trap question because you don't know. A better question is: *How much drawdown could I emotionally tolerate without changing my plan?*

Try this thought experiment. Imagine you invest the entire lump sum today. Over the next month, it drops 10%, 20%, 30%. At what point do you feel a strong urge to intervene—sell, stop contributing, move to cash, or "wait for things to stabilize"?

That number is your *panic threshold*. It's not a confession of weakness. It's a measurement, like knowing your max heart rate before designing a workout plan. If you know you would capitulate at a 20% drop, then a lump-sum strategy is only "optimal" in the same way that sprinting a marathon is optimal: it looks great for the first mile.

DCA lowers the chance that you experience an immediate, identity-threatening loss on the full amount. It doesn't remove risk. It changes the distribution of emotional outcomes.

If your panic threshold is high—you can truly shrug at volatility and keep buying—lump sum is often appropriate. If your panic threshold is low, DCA can be a rational concession to how humans actually behave.

### The silent killer is not choosing wrong; it's waiting without a rule

The most common failure mode with a lump sum isn't investing it and immediately losing money. It's never investing it at all.

Waiting feels prudent because it doesn't create a loss on paper. But "no loss" is not the same as "no cost." The cost is time. And in accumulation, time is the raw material of compounding. Uninvested cash can become a lifestyle. One month becomes six. Six becomes a year. Then you wake up to a market that has already moved, and now you're scared to invest because it *already went up* and you missed it. The mind is spectacularly creative at finding reasons to delay both before and after an event.

This is where DCA shines, even for investors who know the math often favors lump sum. DCA is a rule that breaks the waiting loop. It gives your brain a job it can approve of: "We are being careful." Meanwhile, your capital actually enters the market.

### DCA isn't timid if it's fast

Many people equate DCA with multi-year hesitation. That's not DCA; that's fear with a calendar.

A practical DCA schedule for a true lump sum is typically measured in months, not years. Think 3, 6, or 12 months as a psychological ramp. The longer you stretch it, the more you're transforming a regret hedge into a return drag. You're also increasing the risk that life intervenes and the plan dies halfway through. A one-year DCA schedule can be a clean compromise: it gives you enough time to feel multiple market moods without turning "careful" into "paralyzed."

The key is to make the schedule *mechanical*. You want fewer moments where you re-open the question. A plan that forces you to decide twelve times is not a plan; it's twelve opportunities to back out.

Automation is not a convenience feature. It's behavior engineering. If your DCA schedule requires you to manually place trades every month, you are asking willpower to do the job of a calendar. Willpower is expensive and inconsistent. Calendars are cheap and relentless.

### Two stories, same market: why DCA can feel like relief even when it ``underperforms"

Suppose markets rise steadily for the next year. Lump sum wins. DCA "underperforms" because some cash sat on the sidelines.

But ask yourself what the "underperformance" bought. For many investors, it bought the confidence to hold stocks at all. It bought the experience of watching the market rise and fall while they continued buying. It bought practice. That's not fluff; it's training for the moment when the market eventually drops hard and headlines get theatrical.

Now suppose markets fall sharply right after you invest. Lump sum loses more initially. DCA loses less at first and then buys more shares at lower prices. Depending on the exact path, DCA may even end up with a higher share count. But the most important difference is psychological: DCA investors are less likely to see the initial drop as a personal catastrophe. They're more likely to keep going, which is the only behavior that matters.

This is why "lump sum wins most of the time" is an incomplete statement. The times it loses are exactly the times you learn whether you can keep your hands off the steering wheel.

### When lump sum is psychologically easier (yes, sometimes it is)

It's tempting to declare DCA the "behavioral" solution for everyone. Not true. For some personalities, DCA is torture because it extends uncertainty. These investors don't fear a single bad entry point as much as they hate the drip-drip of second-guessing. They'd rather rip off the bandage, invest once, and move on with their lives.

If you know you'll obsess every month—watching markets, trying to "optimize" the next tranche, re-litigating the schedule—then lump sum can be the behavioral choice. One decision. One execution. Then your attention can return to your career, your family, and the much more important task of continuing to add money over time.

This is the overlooked truth: the best accumulation strategy is the one that *protects your attention*. Attention is a scarce asset, and the market will happily consume it if you let it.

### A practical framework: split the problem into ``risk capacity" and ``regret capacity"

Traditional finance talks about risk tolerance: how much volatility you can stomach. That's useful, but it's incomplete because "stomach" sounds like mood. Accumulation needs two capacities:

*Risk capacity* is your financial ability to withstand losses. If you have a stable job, a long time horizon, and an emergency fund, your risk capacity is often higher than you feel.

*Regret capacity* is your psychological ability to withstand feeling wrong. This is the part people hide from themselves. Some investors can watch a portfolio drop 30% and keep buying because they interpret it as a sale. Others cannot, even if they can afford it financially, because the emotional experience is too corrosive.

Lump sum is primarily a bet on risk capacity (and expected return). DCA is primarily a tool for regret capacity (and follow-through).

To use this framework honestly, you don't ask, "Am I brave?" You ask, "Which regret would I rather live with?" Because you will live with one of them. Investing is a regret-allocation problem.

### Regret minimization as a rule: choose the strategy that prevents the worst behavioral outcome

Here is the regret-minimization decision rule that works in the real world:

> Choose the strategy that makes the probability of *not investing* as close to zero as possible.

If you lump-sum invest, will you be tempted to bail after a drop? If you DCA, will you be tempted to pause the plan? If either temptation is strong, the strategy isn't "wrong," but it is incomplete. It needs guardrails.

Guardrails can be simple:

- Pre-commit to a DCA schedule and automate it.
- If lump-summing, pre-commit to a "no selling for X years" policy for the equity portion.
- Separate the money mentally: some for near-term safety, some for long-term growth, so you're not emotionally forcing every dollar to do every job.

That last point matters. A common psychological reason people DCA is that they're trying to use one pile of money for two contradictory goals: long-term growth *and* short-term emotional safety. If you don't explicitly allocate for safety elsewhere (emergency fund, near-term needs), your growth portfolio becomes the hostage of your anxiety.

### Make the decision once; stop re-arguing it with yourself

The market will tempt you to re-open the debate every time it moves. If you DCA and the market jumps, you'll feel pressure to speed up. If you lump sum and the market drops, you'll feel pressure to "wait until it comes back" (which is just a fancy way of saying "sell low mentally and buy high later").

The way out is to treat the decision like a policy, not a prediction. Policies don't require daily reinforcement. They require occasional review.

If you want a concrete policy template, here's one that many serious investors can live with:

> If I receive a lump sum, I will invest it according to my target allocation either immediately or over a fixed 6–12 month automated schedule. I will not change the schedule based on news, feelings, or market levels.

It reads almost boringly simple. That's the point. Accumulation isn't won by brilliant tactics. It's won by removing the number of times you have to be brilliant.

### Actionable takeaway

Write down your lump-sum deployment policy *today*, while you're calm: choose either (a) invest immediately, or (b) a fixed automated DCA schedule of 3–12 months. Then add one sentence that makes it mechanically enforceable: *"I will not pause, accelerate, or delay based on market moves or headlines."* The goal isn't to pick the mathematically perfect entry point; it's to make sure the money actually gets in—and stays in long enough for compounding to do its quiet, inevitable work.

## 6.2 The Transition: Human Capital and Glide Paths

At 25, your biggest asset probably doesn't have a ticker symbol. It's your ability to wake up on a random Tuesday and turn time into money. That sounds obvious until you realize what it implies: when you're young, your net worth statement is lying by omission. It lists the checking account, the student loans, maybe a small 401(k). It usually doesn't list the thing doing all the heavy lifting—the future paychecks that haven't arrived yet.

That invisible asset is *human capital*: the present value of your future earning power. Present value just means "if you could convert all those future paychecks into one lump sum today, what would that lump sum be worth?" It's a strange concept the first time you see it, and then it becomes impossible to unsee. A 28-year-old engineer with $30,000 invested and a stable career might have a small portfolio but enormous human capital. A 62-year-old with $2 million invested may have the opposite: massive financial capital and very little human capital left.

The key transition in adult finance is not when you hit a round-number birthday. It's when the *composition* of your wealth flips. Early on, you're mostly a "paycheck machine" with a small investment account attached. Later, you're mostly an "investment machine" with a paycheck attached. The portfolio stops being a sidecar and starts becoming the engine.

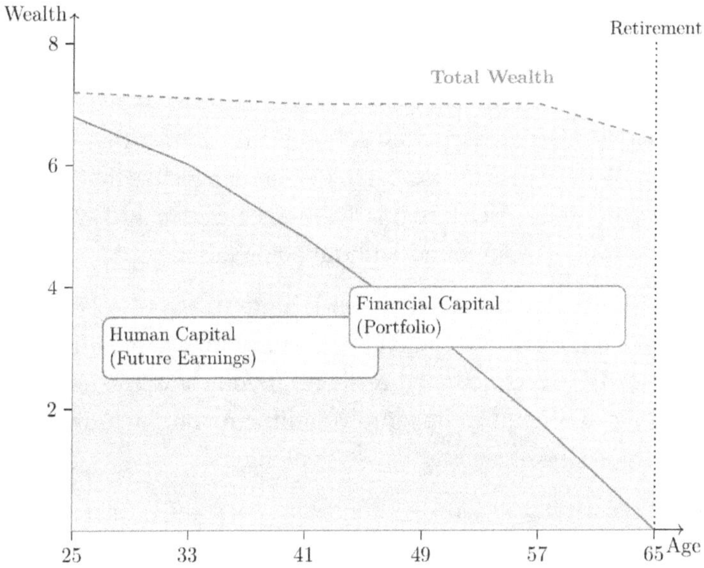


If that picture feels a little provocative, good. It's meant to be. It suggests something that clashes with the way most people talk about investing: your stock/bond mix isn't just a preference. It's a *replacement strategy*. As human capital drains away, the portfolio has to take over the job of stability.

That's where glide paths come in. A glide path is a planned change in your asset allocation over time, usually shifting from riskier assets (like stocks) toward steadier ones (like bonds and cash) as retirement approaches. People often treat glide paths as etiquette: "responsible adults own more bonds." The deeper reason is mechanical. When you're young, the *paycheck* is the stabilizer. When you're older, the *portfolio* must become the stabilizer. You don't de-risk because you become timid; you de-risk because the shock absorber is being removed.

### Why human capital behaves suspiciously like a bond (until it doesn't)

Calling human capital "bond-like" sounds like academic wordplay, but it's a useful mental model. A bond is a contract that pays you fairly predictable cash flows. A stable salary is a lot like that: a stream of payments that arrives regularly, financing your life and letting your investments compound without being tapped.

In that sense, a tenured professor, a long-time civil servant, or a nurse in a high-demand region may have human capital that's relatively steady, diversified, and resilient. If your paycheck is dependable, you can often tolerate more volatility in your investment portfolio because the *rest of your life* isn't volatile.

But human capital isn't always bond-like. For some careers it's closer to a leveraged stock position. If you work in a cyclical industry, rely on bonuses, hold a lot of company equity, or your job security is tied to the market's mood, your future earnings may drop at the same time stocks drop. That correlation is a big deal. It means the moment your portfolio gets punched in the face is also the moment your income may wobble. That is the financial version of slipping on ice while carrying groceries.

So human capital has two traits that matter for asset allocation:

- *Size*: how much future income you still have ahead of you.
- *Stability*: how "bond-like" that future income is—and how correlated it is with the market.

A large, stable human capital base can justify an aggressive portfolio because your total wealth is already anchored. A large but fragile human capital base is a warning label: you may need portfolio stability *earlier* than your age suggests.

### 6.2.1 The brutal arithmetic of time: human capital is a melting asset

Every year that passes, you don't just get older. You consume an irrecoverable resource: remaining working years. You can't "catch up" on a missing decade of earnings the way you can catch up on a missed workout. Even if you earn more later, you can't rewind the calendar and make those dollars start compounding earlier.

That's why human capital is a melting asset. The balance declines as the years of potential earnings decline. In the early years, it can feel infinite. Then it suddenly isn't. Late-career people often describe a strange psychological shift: the paychecks become precious not because they're higher, but because there are fewer left.

This melting quality changes what risk means. When you're 30, a market crash is frightening but also, in a dark way, *useful*. You're buying future returns with future contributions. When you're 60, a crash is the same price movement, but it lands on a much larger financial base and a much smaller future contribution stream. The math becomes less forgiving because your ability to refill the tank is weaker.

### 6.2.2 Glide paths: not an age rule, a replacement schedule

A glide path done well is not "more bonds because you're older." It's "more safe cash flows because your paycheck is fading."

Imagine two 45-year-olds:

- Person A has a stable job, strong benefits, low probability of unemployment, and a high savings rate.
- Person B is self-employed with volatile income and no real safety net beyond their portfolio.

They are the same age, but their *total wealth structure* is different. Person A's human capital is sturdier. Person B's is more fragile. If both hold the same aggressive portfolio because a birthday told them to, one of them is taking a risk they don't understand.

Glide paths exist to manage that misunderstanding. They're a way to gradually convert "future paychecks" into "assets that throw off reliable return streams." You're swapping what finances your life from one engine to another. The smoother the handoff, the less likely you are to panic at exactly the wrong moment.

### 6.2.3 Target-date funds: the default glide path you probably own without realizing it

Target-date funds (TDFs) are the mass-market version of a glide path. You pick a year (often near your planned retirement date), and the fund automatically adjusts its mix of stocks and bonds over time.

They're popular for a reason: they turn a complicated ongoing decision into a one-time choice. That's not laziness; it's good system design. If you're the kind of investor who will otherwise tinker yourself into trouble, a sensible default can be better than a bespoke plan you won't stick to.

But TDFs come with a subtle trap: two funds with the same target year can be meaningfully different. The glide path can be more aggressive or more conservative. Fees can differ. The stock/bond split at retirement can differ. A label is not a guarantee; it's a category.

One particularly important distinction is whether a fund is designed "to" retirement or "through" retirement. A "to" design often reaches its most conservative mix around the target date. A "through" design may remain more equity-heavy at retirement and continue de-risking after. The names sound like marketing, but the difference is real: it changes how much stock exposure you'll carry exactly when the portfolio stops being a background character and becomes your co-pilot.

### 6.2.4 The uncomfortable question: what if your human capital is already ``retiring" early?

Not all transitions happen on schedule. Some happen because life forces them: health issues, caregiving, layoffs, burnout, or an industry that changes faster than a career can adapt. Human capital can drop not gradually, but suddenly.

This is why treating glide paths as age-only is dangerous. A 50-year-old with a high-stress job and worsening health may have less human capital than a 60-year-old who loves their work and could easily consult part-time. Flexibility is part of human capital. The ability to work longer, reduce hours, or switch roles without blowing up your income is a form of resilience. It's a financial asset disguised as a lifestyle choice.

A portfolio that assumes you'll work until 67 is fragile if you're secretly hoping you can't. Your asset allocation shouldn't be a fantasy of future you; it should be a plan that survives future you having a bad year.

### 6.2.5 A glide path is also a story you tell yourself during market stress

Asset allocation is math, but it's also narrative. A good glide path gives you a calm explanation when markets get loud:

> "I own bonds not because I'm pessimistic. I own bonds because I'm replacing the stability my paycheck used to provide."

That story matters because it prevents a common midlife mistake: becoming risk-averse *in the wrong way*. Many investors reduce stock exposure after a bear market, not as a planned transition, but as a reaction to pain. That's not de-risking; it's flinching. A glide path tries to move risk *before* markets force your hand, using a schedule rather than a headline.

### 6.2.6 The ``bond-like paycheck" can justify stocks---but only if you also manage the other risks of being human

Here's where the human capital lens gets refreshingly honest. If your paycheck is the stabilizer, then threats to the paycheck are threats to your total wealth. Managing those threats is part of asset allocation, even if it doesn't look like investing.

If you're early in your career and heavily invested in stocks, your real risk management might be:

- keeping a real emergency fund (so a layoff doesn't force liquidation),
- carrying appropriate insurance (so an injury doesn't vaporize earning power),
- avoiding a lifestyle that requires stock sales to fund basic bills.

None of that appears on a brokerage statement, but it changes how much risk your portfolio can safely take. A young investor with no cash buffer and aggressive stocks isn't "brave." They're illiquid.

That's the deeper point: human capital and financial capital are a single system. If you treat them as separate, you'll be surprised at exactly the wrong time.

### 6.2.7 When the classic glide path logic gets controversial

Some research and practitioner debates argue that the riskiest period isn't the years leading up to retirement; it's the early years *after* retirement, when withdrawals begin and the portfolio is most vulnerable to bad sequences of returns. That idea can lead to "rising equity glide paths," where equity exposure starts lower at retirement and increases later.

Whether you adopt that design or not, it highlights something important: glide paths are not commandments. They're risk-management hypotheses. The right glide path depends on:

- how flexible your spending is,
- whether you have pensions or other guaranteed income,
- whether you plan to keep working part-time,
- how you react emotionally to drawdowns.

A glide path is supposed to make your financial life *more survivable*, not more orthodox.

### 6.2.8 A simple way to think about your own glide path without a PhD

You can do an unusually high-quality version of this thinking with two questions.

First: *How bond-like is my income?* Stable employment with predictable raises is bond-like. Commission-only sales, startup equity dependence, and cyclical industries are less so.

Second: *How many paychecks do I have left that matter?* Not "until retirement age," but until you could realistically stop working without your lifestyle collapsing. If that number is big, your human capital is still doing most of the stabilizing work. If that number is small, your portfolio is already auditioning for the lead role.

Then build a glide path that reflects those answers. A stable-income 35-year-old may rationally hold a high equity allocation, especially if they can keep contributing through downturns. A 45-year-old with volatile income and a thin safety buffer may want more ballast than the typical age rule suggests, because their human capital is equity-like and their ability to "just keep buying" is conditional.

#### Actionable takeaway

Write a one-paragraph "total wealth" memo and use it to sanity-check your asset allocation:

> (1) My human capital is approximately ___ (stable / somewhat volatile / highly volatile) and is ___ (positively / weakly / negatively) correlated with stock markets.
> (2) If I lost my job or had to reduce work for 12 months, my portfolio plan would ___ (still be intact / require selling / collapse).
> (3) Therefore, my glide path rule is: each year (or every two years) I will shift ___ percentage points from risky assets to safer assets *unless* my human capital remains unusually stable and my safety buffer is strong.

It's not the percentage point that does the magic. It's the habit of treating your paycheck as part of the portfolio. When you can see human capital clearly, de-risking stops feeling like surrender and starts feeling like what it is: a planned handoff from one compounding machine to another.

## 6.3 The Danger Zone: Sequence of Returns Risk

Two retirees can own the same portfolio, earn the same *average* return, and still end up in different financial universes. One spends freely, tips generously, and dies with a portfolio that kept growing long after it "should have" settled down. The other spends the first decade of retirement doing math in their head at the grocery store, not because they were reckless, but because the market happened to be rude *early*.

That difference has a name: *sequence of returns risk*. It's the risk that the *order* of market returns matters more than the average return, specifically when you're withdrawing money. During accumulation, a bad early market can feel like a discount. During retirement, a bad early market can be a structural injury.

The cruel part is that sequence risk doesn't announce itself as a special kind of risk. It masquerades as ordinary volatility. The S&P 500 drops, the news gets dramatic, your portfolio goes down. Same movie you've seen before. The difference is that, now, you're not only watching the numbers move. You're *removing pieces* of the machine while it's running.

### Why averages stop being comforting the moment you start withdrawals

In accumulation, the logic of compounding is friendly: money stays in the system, and new money keeps entering. You can survive ugly years because you're still buying, still earning, still adding fuel. In retirement, you've switched from contributing to consuming. The portfolio is no longer just compounding; it's compounding *and* being harvested.

That harvesting is what breaks the "average return" intuition.

A simple way to see it is to separate two questions that get bundled together:

- *What did the market do?* (returns)
- *What did you do while the market did that?* (cash flows)

Once you're withdrawing, your cash flow is negative. You are selling shares (directly or indirectly) to fund life. If the market drops early, you must sell *more shares* to raise the same number of dollars. Those shares don't get to participate in the eventual recovery because they're gone. The recovery happens, but it happens on a smaller base. That's the failing mechanism in one sentence.

This is why sequence risk is sometimes described as "reverse dollar-cost averaging." When you were accumulating, you were happy to buy more shares when prices were low. In retirement, you're forced to *sell* more shares when prices are low. Same arithmetic, opposite sign, completely different emotional experience.

## The anomaly: same returns, different order, different destiny

The cleanest demonstration uses a slightly unsettling trick: give two retirees the exact same set of annual returns, just in a different order. Same multiset of returns. Same "average." Same long-run market environment. The only difference is timing.

Then add one more realistic ingredient: a steady withdrawal for spending.

One retiree gets a lucky sequence: strong returns early, weak returns later. The other gets the unlucky sequence: weak returns early, strong returns later. The lucky retiree's portfolio grows early, so later withdrawals are small relative to the remaining wealth. The unlucky retiree's portfolio shrinks early, so withdrawals become a larger and larger percentage of what remains. Eventually the withdrawals stop being spending; they become demolition.

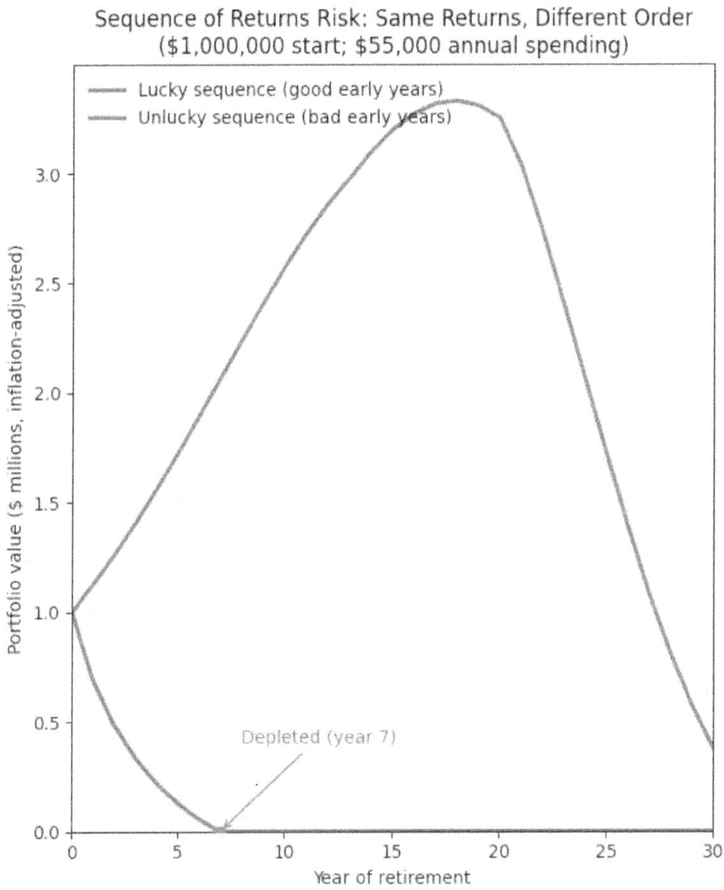


This picture is the reason retirees get whiplash when someone says, "The market averages about X%." The average is a property of the market. Retirement success is a property of the *interaction* between the market and your withdrawals.

And no, this isn't a cherry-picked academic stunt. It's a model of a very human experience: retiring into a bad decade. It doesn't require a once-in-a-century depression. It only requires that your first handful of years are unkind while you're extracting cash and not adding new money.

## The failing mechanism: why early losses are different losses

Losing 20% in year 1 of retirement is not "the same" as losing 20% in year 20.

In year 1, your balance is typically at its lifetime peak (you've spent decades accumulating), and your spending is just beginning. If the portfolio drops, you still need to pay property taxes, buy groceries, help a kid with a wedding, fix the roof. Spending doesn't politely pause for a bear market. So you sell into weakness.

Selling into weakness does two kinds of damage:

- *Mathematical damage*: you permanently reduce the number of shares left to recover.
- *Psychological damage*: you begin retirement with a story of fragility, which nudges you toward either panic selling or chronic under-spending.

In year 20, even if the market drops, the portfolio may have already grown enough that your withdrawal is a smaller fraction of the total. You have a bigger cushion. You might also have a more flexible life: maybe the mortgage is gone, maybe Social Security is covering a larger share, maybe you're simply less interested in expensive consumption. The portfolio can take a hit without forcing the same "sell low" mechanics.

Sequence risk is why retirement math is full of people who did everything "right" and still got unlucky. It's also why a retiree who looks rich on paper can still feel poor: their brain is reacting not to their average return, but to the path their money took.

## The hidden lever: withdrawal rate is not a number, it's a force

A withdrawal rate is often presented like a static fact: "Take 4% and adjust for inflation." But in the real system, a withdrawal is a force applied every year regardless of weather.

When markets are up, the withdrawal is gentle. It's a small slice of a growing pie.—When markets are down, the same dollar withdrawal is aggressive. It becomes a larger slice of a shrinking pie.

This is why two retirees can start with the same "safe" plan and experience it differently. The plan isn't just a rate; it's a rate *relative to current portfolio value*. In bad sequences, that relative rate spikes, even if spending never changes.

If you want a single sentence to remember, make it this:

> In retirement, returns are multiplied by time, but withdrawals are multiplied by *bad timing*.

### Why this risk shows up right when you're least interested in drama

There's a particular insult baked into sequence risk: it tends to matter most when you're least emotionally equipped for it.

Early retirement is a psychologically charged phase. You've just stepped away from the income stream that used to patch mistakes. Your portfolio is suddenly responsible for groceries, not just "future you." And you're trying to calibrate a new identity: not worker, not saver, but spender-who-must-not-run-out.

Then the market does what it always does: it occasionally behaves like a maniac.

When that happens in the first few years, retirees often respond with one of two extreme behaviors:

- *Panic protection*: slash spending so hard that retirement becomes a long apology.
- *Denial*: keep spending as if nothing changed, hoping the market will "come back" before the math catches up.

Neither response is irrational emotionally. Both can be costly financially. The real win is having a plan that makes neither response necessary.

### What actually reduces sequence risk (without pretending you can predict markets)

There are only a few levers that reliably matter, and they're refreshingly unsexy.

*Spending flexibility* is the biggest one. The more your spending can adjust, the less the portfolio has to. That doesn't mean living in fear; it means distinguishing between spending that is identity-critical (housing, healthcare, core lifestyle) and spending that is timing-optional (travel, gifts, renovations, new car cadence). A plan that can temporarily dial down optional spending in bad markets is structurally stronger than a plan that treats every dollar as sacred and fixed.

*A cash buffer* can help because it gives you time. Time is what you need in the exact moment sequence risk bites. If you can fund spending from cash or short-term bonds while equities recover, you avoid forced selling at depressed prices. This isn't a claim that "cash is great." It's a claim that cash can be a behavioral and mechanical bridge when your system is most fragile.

*Asset allocation* matters, but not in the simplistic way people hope. Holding fewer stocks can reduce volatility, which can reduce the chance of a catastrophic early drawdown. But too little equity can introduce a different risk: not earning enough long-run growth to sustain decades of withdrawals. The point isn't "more bonds" or "more stocks." The point is aligning your allocation with your ability to avoid selling equities at the worst time.

*Rebalancing discipline* is another quiet lever. In retirement, rebalancing isn't about squeezing out a tiny return edge. It's about controlling what you sell. If stocks fall and bonds hold up, a rebalancing-aware withdrawal strategy may pull more from the bond side, letting equities heal. If you withdraw "pro-rata from everything," you may unintentionally lock in the very losses you can't afford.

And, most importantly, *withdrawal rules* can be designed to respond to sequence risk. A rigid rule says, "I take this amount no matter what." A dynamic rule says, "I take this amount, but I have pre-committed adjustments if the portfolio is stressed." That trade—some lifestyle variability in exchange for durability—is often the most honest form of safety.

### The emotional reframing that keeps retirees from self-sabotage

Sequence risk is sometimes taught like a monster hiding under the bed. That can backfire. If retirees feel like the first bad year proves their plan is doomed, they'll do something extreme.

A better framing is to treat early retirement as a stress test that your plan expects.

If you get bad early returns, it doesn't mean you made a bad plan. It means you drew a hard card first. The response isn't to "fix the portfolio" in a panic. The response is to execute the pre-written adjustments: temporary spending trims, a cash bucket drawdown, disciplined rebalancing, maybe part-time income if that's in the range of possible.

The goal is not to avoid discomfort. The goal is to avoid *irreversible actions* taken during discomfort.

### Actionable takeaway

Run a simple sequence-risk drill on your retirement plan: assume the first two years of retirement are down 25% and down 15% (in any order), while you keep withdrawing your planned spending. Write down, in advance, what you will do *mechanically* in that scenario:

- the maximum spending cut you could tolerate for 12–24 months,
- which assets you would sell first to fund withdrawals (cash/bonds before equities, if available),

- the one action you forbid yourself from taking (for many people: selling all equities).

If you can't write a credible answer, that's not a moral failure. It's valuable feedback: your withdrawal plan is relying on luck. Sequence of returns risk is the part of retirement math that punishes luck dependence.

## 6.4 Decumulation: From Safe Rates to Dynamic Rules

The most expensive sentence in retirement is usually spoken with confidence: *"We can spend $X per year. It's safe."* The number feels like a bridge from uncertainty to calm. It turns a volatile portfolio into a paycheck substitute. It lets you book the trip, remodel the kitchen, and stop refreshing your account balance like it's a blood pressure monitor.

And that's exactly why people love the 4% Rule. It offers a clean, memorable deal: withdraw 4% of your starting portfolio in year one, then raise that dollar amount with inflation every year, and historically you would have survived most 30-year periods with a stock-/bond portfolio. It's the kind of rule that fits on a cocktail napkin, which is both its genius and its danger.

Because a cocktail napkin can't hold the real problem: retirement spending is not a one-time calculation. It's a *control system*. Markets move. Inflation moves. Your health moves. Your desires move. A rigid withdrawal rule tries to freeze a living thing in amber.

Dynamic rules do something more honest. They admit that the portfolio isn't a pension. It's a machine that sometimes runs hot and sometimes sputters. Your job in decumulation is not to find the perfect constant number; it's to design a spending process that keeps the machine alive *and* keeps your life worth living.

### Why the 4% Rule became famous (and why it still matters)

It's fashionable to dunk on the 4% Rule, but it earned its fame. It taught a generation of investors a crucial lesson: the portfolio doesn't need dividends or interest to "fund retirement." You can sell shares. You can live off total return. And you can think probabilistically rather than superstitiously.

It also anchored a conversation that used to be pure vibes. Before safe withdrawal research went mainstream, retirees often followed rules like "never touch principal" or "only spend the interest," which sounds conservative until you realize it can force absurd underspending and unnecessary risk-taking for yield.

So keep the 4% Rule in its rightful place: not as a prophecy, but as a baseline. It's the Model T of retirement income planning—not luxurious, not fast, but it changed the world by proving the thing could work at all.

Then remember what the Model T didn't have: airbags, traction control, crumple zones. Dynamic rules are the safety systems.

### The problem with ``safe": it protects money by starving life

Here's the psychological trap: a fixed real withdrawal is designed to be robust to terrible markets. That robustness often requires planning for worst-case history. Planning for worst-case history means you may spend your entire retirement living as if disaster is always a month away.

If markets cooperate, the "safe" plan can create a weird outcome: you die with a mountain of money and a lifetime of postponed joy. The portfolio did fine. The life plan was overly cautious. A rule that guarantees you won't run out can still fail if it guarantees you won't fully live.

This isn't an argument for recklessness. It's an argument for alignment. The point of money is not to maximize the probability of leaving an estate; it's to fund a life under uncertainty. Some people deeply want to leave a large legacy. Others would rather spend more early while their knees still work and their friends are still alive. A rigid rule can accidentally impose someone else's values on your calendar.

Dynamic spending rules are, at their best, values-aware. They can allow higher spending when the portfolio is healthy and gently apply brakes when the portfolio is stressed. Instead of one constant number, you get a *range* of sustainable living.

### Sequence risk is the villain; rigidity is the accomplice

The danger isn't that retirees spend too much in random years. It's that they spend too much *in the wrong years*—specifically early in retirement after a market drawdown, when sequence-of-returns risk is most lethal.

A fixed inflation-adjusted withdrawal doesn't care what the market is doing. It keeps raising the dollar amount as if the portfolio were a salary with cost-of-living adjustments. That's the rigidity. When early returns are poor, rigidity forces you to sell more shares at low prices, accelerating the very mechanism that can break the plan.

Dynamic rules are essentially anti-rigidity devices. They don't try to eliminate sequence risk (you can't). They try to stop you from feeding it.

### Two families of dynamic rules: guardrails and percentages

Most dynamic strategies belong to one of two families. Both are defensible. They just optimize for different kinds of comfort.

*Guardrails* keep spending relatively stable most of the time, but they include explicit triggers that cause adjustments. Think of them like a thermostat: the temperature is allowed to drift inside a comfort band, and only when it hits the edge do you hear the system click on.

A simple guardrail approach might look like this in plain language:

> "I'll start with a planned spending level. If the portfolio does very well, I give myself a raise. If the portfolio does very poorly, I temporarily tighten the belt. Otherwise, I keep spending roughly steady."

The power is not in the precise thresholds; it's in the pre-commitment. You are deciding your future behavior *now*, when fear isn't driving.

*Percentage-of-portfolio rules* (including approaches like Variable Percentage Withdrawal, or VPW) are even more direct: each year you spend a chosen percentage of whatever the portfolio is worth. If the portfolio drops, your spending drops automatically. If it rises, your spending rises. This method is brutally sustainable because it tends to avoid depletion by construction: you can't spend money that isn't there.

The tradeoff is lifestyle volatility. A percentage rule can produce spending swings that feel unacceptable if you need a steady check for fixed expenses. It's a great rule if you have other stable income (like Social Security, a pension, or part-time work) covering your non-negotiables and the portfolio is funding the flex.

If guardrails are a thermostat, percentage rules are a dimmer switch wired directly to the market.

### The underrated innovation: separating ``floor" spending from ``flex" spending

Retirees often argue about withdrawal rules as if every dollar of spending is equally important. It isn't. Some spending is load-bearing: housing, utilities, basic food, insurance, healthcare, taxes. Some spending is the good stuff that makes retirement feel like freedom: travel, gifts, hobbies, restaurants, home projects, helping family.

Dynamic spending becomes emotionally easier when you separate those two categories.

If your "floor" is funded by stable sources—Social Security, pension, annuity income, or a conservative bond ladder/cash reserve—then your portfolio withdrawals can be more dynamic without threatening your sense of safety. You're no longer asking the market to pay the electric bill on schedule. You're asking it to fund the fun, which can be timed.

This separation also makes the guardrails conversation less scary. A retiree who hears "spending cut" imagines losing the house. A retiree who hears "travel budget cut for 18 months" imagines disappointment, not catastrophe. Same math, different nervous system response.

### A worked example of the psychology (not the spreadsheet)

Imagine a couple retiring with $1.5 million. They want to spend $75,000 from the portfolio plus Social Security later. Their first year goes fine. Then the market drops 25%. Their portfolio is down, but their withdrawal plan is still calling for the same inflation-adjusted spending. They are now being asked to do something unnatural: sell more shares in a year that already feels like failure.

This is the moment many retirees discover whether their plan is a plan or a hope.

A guardrails approach changes the conversation. It says: "We anticipated this. When the portfolio is stressed, we skip the inflation raise and trim discretionary spending by $5,000–$10,000 for now." It's not austerity; it's tactical patience. It buys time for the portfolio to recover without permanent share-count damage.

A percentage approach changes the conversation even more. It says: "This year's spending is 4.5% of current balance. So we spend less this year. Automatically." The couple may not love it, but the system is coherent. They're no longer arguing with the market. They're responding to it.

The real point: dynamic rules aren't about optimizing returns. They're about preventing the portfolio from being forced into the one behavior it cannot afford: selling a lot after a big drop.

### Dynamic rules also solve a different retirement problem: the ``permission to spend" problem

Not all retirement risk is running out of money. There's also the risk of never using the money.

Many retirees experience an emotional lag. They spent decades saving, optimizing, delaying gratification, getting a small dopamine hit from watching balances rise. Then retirement arrives, and they can't flip the switch. Spending feels like breaking something sacred. They keep living like they're "almost there," even when they're there.

A rigid safe withdrawal number doesn't always fix this. Sometimes it makes it worse: the number feels like a maximum, not a plan. Every purchase becomes a debate: "Is this within the 4%?" The rule that was supposed to create freedom becomes a ceiling that creates guilt.

Dynamic rules can help because they provide *conditional permission*. You aren't just allowed to spend; you're allowed to spend more when the portfolio is healthy. That turns spending increases into something like rebalancing: a mechanical response rather than a moral failing.

A raise becomes part of the policy. That matters more than most spreadsheets admit.

### The real cost of flexibility: it must be negotiated in advance

Flexibility is easy to recommend and hard to live.

If a plan says "cut spending when the market is down," a retiree may nod in January and revolt in October. Not because they're irrational, but because "down" feels personal when it's your life funding the drawdown.

That's why dynamic rules work best when they're explicit. Vague flexibility is just anxiety. Specific flexibility is a plan.

A good dynamic rule answers uncomfortable questions ahead of time:

- How much can spending drop in a bad year without harming dignity?
- Which categories get cut first?
- What portfolio condition triggers a raise or a cut?
- What is the maximum cut you're willing to tolerate, and for how long?

You're not predicting the future. You're writing a response protocol so your future self doesn't have to improvise under stress.

### How to think about ``failure" when rules are dynamic

With a fixed rule, failure is simple: you ran out of money, or you didn't. With dynamic rules, "failure" becomes more nuanced. You might never deplete the portfolio, but you might experience spending volatility that feels like failure if your lifestyle requires stability.

So evaluate the strategy the way you evaluate a good insurance policy: not only by whether it prevents ruin, but by what it costs you in normal times. Guardrails tend to cost less lifestyle stability in normal markets, but they can demand decisive cuts in bad markets. Percentage rules cost more stability year to year, but they can be remarkably robust against depletion.

Neither is "best." The question is what kind of discomfort you're willing to tolerate:

- discomfort from occasional planned cuts (guardrails), or

- discomfort from frequent variability (percentage-based).

Retirement planning is an exercise in choosing which discomfort you prefer to pre-pay for.

### A practical synthesis: the hybrid rule many retirees actually live with

In real life, many retirees end up with a hybrid whether they name it or not:

- A baseline spending level that covers needs (supported by stable income and conservative assets),
- plus a variable "bonus" budget that expands and contracts with the portfolio.

This hybrid has an emotional advantage: it allows the retiree to feel both safe and alive. The floor gives dignity. The variable layer gives upside participation without threatening survival.

It also respects a truth that gets lost in debates: retirement isn't one long uniform season. The early years tend to be more active (and often more expensive). Later years can become naturally less spendy, or they can become more expensive due to healthcare. A dynamic rule can accommodate changing realities without demanding that you pretend you'll want the same lifestyle at 85 that you want at 65.

### Actionable takeaway

Write your withdrawal policy as a simple *if/then* script and pin it to your retirement plan:

> *Base spending*: $___ per year (non-negotiables), funded by Social Security/pension/cash/bonds.
> *Portfolio spending*: $___ in year one.
> *Guardrails*: If the portfolio (after withdrawals) is down more than ___% from its starting value (or from its prior peak), then next year I (a) skip the inflation increase and

(b) cut discretionary spending by $___ until recovery.
*Raises*: If the portfolio is up more than ___% above plan, then I increase discretionary spending by $___ (or by ___%).
*Non-negotiable rule*: I do not increase withdrawals after a down year.

The exact thresholds can be customized later. The big win is deciding in advance that retirement spending will adapt to markets rather than pretend markets will adapt to your spending. That one shift—from a rigid "safe rate" to a living rule—is often the difference between a portfolio that survives and a retiree who feels like they can breathe.

## 6.5 Cash Management: The Psychological Buffer

Cash has an image problem. In investing circles it's treated like the embarrassing friend who never went anywhere: safe, boring, and faintly disappointing. The math people call it "cash drag," as if it's a parachute you insisted on wearing while trying to run a marathon. And in pure long-run expected-return terms, they're not wrong: cash usually yields less than stocks, and often less than inflation, which means it can quietly shrink your purchasing power.

But retirement isn't a long-run expected-return contest. Retirement is a years-long exercise in *not doing something stupid at the worst possible moment*. Cash, used well, is not a return asset. It's a behavior asset. It buys you the ability to look at a collapsing market and say, truthfully, *"We don't have to sell today."*

That sentence is worth more than most people think.

## The real function of cash: preventing forced selling

When markets fall, the damage to a retiree is rarely the drop itself. The drop is common and temporary. The real injury comes from what the drop *forces*: selling shares at depressed prices to fund expenses, locking in losses and reducing the share base that would have recovered. That is the sequence-of-returns mechanism wearing practical clothing: the mortgage company doesn't care that your portfolio is "down but likely to bounce back."

Cash changes the timing. It creates a buffer between "market price today" and "need to pay the bills today." It's the financial equivalent of a storm cellar: you don't move into it permanently, but you're grateful it exists when the weather turns.

This is why thinking of cash as "uninvested" misses the point. Cash is invested in *optionality*. It gives you options when everyone else has none: the option to wait, the option to rebalance rather than liquidate, the option to be patient when patience is most valuable.

## A retiree's nightmare is not volatility; it's a calendar

During accumulation, volatility is an annoyance. You have time, income, and contributions coming in. In retirement, volatility is paired with a calendar full of non-negotiable obligations: quarterly estimated taxes, property taxes, insurance premiums, healthcare bills, the adult kid who needs help *right now*. The market's worst habit is that it does not schedule its bad years around your cash needs.

That calendar turns selling into a compulsion. People often say they "had to sell" during downturns as if they were personally unlucky. Usually they were structurally under-buffered. Their portfolio was designed for average weather, and then it rained.

Cash is how you design for rain without turning the whole portfolio into a puddle.

### Mental accounting: why cash works even when it's not "optimal" on paper

Economists like to remind us that money is fungible. A dollar is a dollar. In practice, human brains refuse to treat dollars as identical. We label money. We build little psychological rooms for it: "emergency fund," "vacation fund," "never touch," "house money." This habit is called *mental accounting*, and it's often mocked as irrational.

In retirement, it can be a superpower.

A cash bucket is a form of intentional mental accounting. You are telling your brain: "This money has one job: keep me from selling stocks in a crash." The label is not a spreadsheet flourish; it's a behavioral guardrail. It creates a clean rule: spending comes from the cash bucket first, not from the volatile bucket. That rule reduces the number of emotional decisions you must make during market stress.

When people argue that "a total-return portfolio makes buckets unnecessary," they're often arguing about economics while ignoring psychology. The bucket strategy isn't only a sourcing method. It's a fear-management method. It is designed for the part of the investor that wakes up at 3 a.m. and negotiates with catastrophe.

### The cash bucket strategy, in plain language

The most common form is simple:

- Keep 1–3 years of spending needs in cash or near-cash (high-yield savings, money market funds, very short-term Treasuries).
- Hold the next layer in high-quality bonds (often intermediate-term).
- Keep the long-term growth layer in diversified equities.

Then follow a simple operational rule:

> Spend from cash. Refill cash when markets are healthy (by selling bonds and/or trimming appreciated stocks via rebalancing). In bad markets, avoid refilling cash from stocks if possible.

Notice what this does: it converts the abstract concept of sequence risk into a practical habit. Instead of saying, "Don't sell low," it creates a system that makes "selling low" less necessary.

The most important word in that description is *refill*. A bucket system that is never refilled is just "holding cash forever," which really can become drag. The system only works if cash is treated as a rotating buffer, not a permanent bunker.

### How much cash is enough? The right answer is behavioral, not universal

A common recommendation is 1–3 years of expenses. That range is useful because it roughly matches the kind of time it often takes for markets to recover from moderate bear markets, and it's long enough to meaningfully reduce panic.

But the exact number should be chosen using a more honest metric: *How many months of spending would let you sleep normally during a 35% equity drawdown?*

Some investors can tolerate a lot of volatility if they have six months of cash because they also have flexible spending and a strong bond allocation. Others, especially those with high fixed expenses or high anxiety about markets, need more than one year for the buffer to do its job. If cash is preventing a catastrophic decision, a little "extra" cash may be cheap insurance.

That said, insurance has premiums. If you hold five or seven years of expenses in cash, you might be paying too much premium: lower long-run growth, higher inflation risk, and the possibility that your "safe" money becomes the thing that makes the plan fragile over decades.

The sweet spot is the smallest buffer that reliably prevents forced selling and panic selling.

### Cash versus bonds: two kinds of safety that behave differently

Retirees sometimes treat "bonds" as the cash bucket. That can work, but bonds are not cash. Bonds can fall when interest rates rise, and longer-duration bonds can be surprisingly volatile. Cash is dull in a special way: it usually doesn't have dramatic drawdowns in nominal terms.

That nominal stability is psychologically powerful. A bond fund might be "safe" in a long-run portfolio sense but still show a scary drawdown right when you wanted it to act like a checking account. If you're building a buffer explicitly to avoid panic, the instrument should behave like a buffer in the moment, not only in theory.

A practical approach many retirees like is a barbell within the safe bucket:

- true cash/near-cash for near-term spending,
- plus high-quality bonds for intermediate resilience and a bit of yield.

The point isn't to maximize return; it's to maximize reliability of the plan under stress.

### The critique: ``Cash drag" is real, and it can quietly hurt sustainability

Now the uncomfortable part. Cash is not free. If you hold a large cash allocation for a long time, you may lower the portfolio's long-run expected return. Over multi-decade retirements, that can matter. Inflation can matter even more. A pile of cash that feels safe can slowly become inadequate.

There's also a structural critique: if you hold a total-return portfolio and rebalance, you may already have a mechanism that sources withdrawals from the "best" place. In a down equity market, bonds may have held up, so withdrawals naturally come from bonds. In a strong equity market, rebalancing trims stocks, refilling the conservative side. In other words, disciplined rebalancing can mimic bucket behavior without explicit buckets.

This critique is not wrong. It's incomplete.

It assumes the retiree will execute the rebalancing plan with calm precision during a bear market. Many will. Many won't. The bucket approach can be thought of as a *training wheels system*: it creates a simpler operational rule that reduces the number of times you have to be heroic.

The decision between "explicit cash bucket" and "rebalance-based withdrawals" isn't purely financial. It's a question about which system you can follow when fear is high and headlines are designed to hijack your attention.

### Cash as a bridge: the practical value of time

The deepest value of a cash buffer is not the dollars. It's the months.

If you have two years of spending in cash, a bear market is no longer an immediate funding crisis. It becomes a waiting problem. Waiting is a dramatically easier problem to solve than "sell now." Waiting allows you to:

- delay large discretionary expenses,
- take part-time income if you choose,
- let dividends and bond interest refill some cash flow,
- rebalance deliberately rather than reactively.

This is why cash is often called "dry powder" in institutional investing. It's not exciting. It's useful *because* it is ready at the exact moment other assets are misbehaving.

### The behavioral failure mode: cash that becomes a hiding place

Cash can also become a place to hide. Some retirees keep increasing their cash allocation after every scary headline. The bucket becomes a bunker. Then they spend from cash for years while refusing to refill it from risk assets, effectively de-risking at the bottom and staying there.

This is the cash version of buying high and selling low. It doesn't look like panic because it's quiet. It looks like "prudence." But it can be devastating over a long retirement.

So a cash buffer needs a refill policy written in ink. The policy can be simple, but it must exist. Otherwise, the buffer can turn into a slow bleed of growth potential.

A good refill policy is not "when I feel like it." It's something like:

> Refill the cash bucket annually if the portfolio is above its prior-year level, or if equities are not in a major drawdown; otherwise delay refilling from equities and spend from bonds/cash.

The key is to pre-commit to what happens after the storm, not only during it.

### Cash is also a social asset

Here's a truth that rarely makes it into retirement spreadsheets: financial decisions are social. If you have a spouse, you don't only need a plan that works. You need a plan that both of you can *believe* in during stress.

A cash buffer can reduce conflict. It can turn "the market is down, should we sell?" into "we have two years of runway." That's a different conversation. It lowers the emotional temperature in the household. That matters, because a plan that causes constant anxiety or arguments is a plan that will eventually be changed—often at the worst time.

The best retirement strategy is one your future self won't sabotage and your household won't revolt against.

### Actionable takeaway

Create a one-page "Cash Buffer Policy" and treat it like a safety manual:

- *Buffer size*: Hold ___ months of core spending in cash/near-cash (pick 12–36 months based on your sleep test).
- *Spending rule*: All monthly spending comes from cash; dividends/interest flow into cash.
- *Refill rule*: Once per year, refill cash back to target *only* if the portfolio is not in a severe drawdown; otherwise fund cash needs by trimming bonds (or delaying discretionary spending).
- *Forbidden move*: Do not increase the cash target in the middle of a market panic.

Cash is not there to make you richer. It's there to keep you from breaking the compounding machine when the market tries to scare you into pressing the wrong buttons. If it lets you hold your long-term assets through a crash without forced selling, it has earned its keep.

## 6.6 Maintenance: The Discipline of Rebalancing

A portfolio without a rebalancing routine is like a house without a thermostat. It might be comfortable for a while. Then one sunny week turns into a heat wave, one cold snap turns into a frozen pipe, and you suddenly discover that "mostly fine" is not a maintenance plan.

Rebalancing is the act of restoring your portfolio to its intended mix of assets—your target allocation. If you decided on 60% stocks and 40% bonds, rebalancing is what you do after the market quietly turns that into 72/28. And the market *will* do that. Left alone, portfolios drift toward whatever has recently been winning. The portfolio becomes a momentum bet without your consent.

That drift feels good while it is happening. It is the financial equivalent of a car that mysteriously "got faster" because gravity helped it roll downhill. People rarely complain about risk creeping up during bull markets. They complain later, when the downhill road ends and the car hits the turn.

#### Risk creep: the stealth tax you don't see on a statement

The most common misunderstanding about rebalancing is that it is primarily about performance. In practice, it is primarily about *risk control.*

When stocks rally for years, your stock allocation becomes a larger fraction of the portfolio even if you never buy another share. That increases expected return a bit, but it also increases volatility and drawdown risk. If you are still accumulating and contributing heavily, that might be tolerable. If you are withdrawing, it can be the setup for a sequence-of-returns problem: you arrive at the vulnerable stage with more equity exposure than you intended.

Rebalancing is how you prevent your own success from rewriting your risk profile.

There's a darker version of risk creep too: it can turn a well-diversified plan into a concentrated bet. A portfolio that started diversified across U.S. stocks, international stocks, and bonds can drift until one asset class dominates. The investor then experiences a bear market not as "markets are down," but as "my portfolio is down," because the portfolio stopped being a portfolio and became a single story.

### The emotional magic trick: rebalancing forces you to do the opposite of your instincts

Humans have terrible instincts at exactly the moments investing asks for good ones. When an asset class has been rising, it feels safe. When it has been falling, it feels dangerous. Rebalancing forces the opposite:

- trim what feels safe (the winner),
- add to what feels dangerous (the loser).

This is why rebalancing is sometimes called "the only free lunch." Not because it guarantees higher returns, but because it systematizes a behavior that most people struggle to do voluntarily: buy low and sell high.

The "free lunch" framing can be misleading, though. Rebalancing is free only in theory. In real life it can create taxes, transaction costs, bid/ask spreads, and behavioral costs (like second-guessing). The goal isn't to rebalance obsessively. The goal is to rebalance *enough* to keep the portfolio behaving like the plan you meant to own.

### The two main rebalancing styles: calendar and bands

Most rebalancing routines come in two flavors.

*Calendar rebalancing* means you rebalance on a schedule: quarterly, semiannually, annually. It is simple, easy to automate, and easy to remember. Its weakness is that it can be either too frequent (creating unnecessary taxes and churn) or too slow (allowing big drift to build up between dates).

*Band (threshold) rebalancing* means you rebalance only when an asset class drifts beyond a certain threshold, such as 5 percentage points away from target or 20% relative drift. For example, if your target is 60% stocks, you rebalance if stocks rise above 65% or fall below 55%. This method is more responsive to volatility and often results in fewer trades in quiet markets.

Band rebalancing also has a psychological advantage: it gives you a reason to act that isn't based on feelings. When your stocks are down and the last thing you want to do is buy more, the band rule turns it into a compliance task: "The allocation breached the policy. We rebalance."

If you've ever watched a disciplined investor calmly buy in a crash and thought they had nerves of steel, there is a good chance you were actually watching a rule do the work.

### The retirement twist: rebalancing becomes your withdrawal engine

In accumulation, contributions can do a lot of rebalancing work. If stocks have run up and are now overweight, you can direct new contributions to bonds or international stocks. If stocks are down, new contributions naturally buy more stocks at lower prices. Your paycheck is a rebalancing tool.

In retirement, that paycheck is gone. But withdrawals can play a similar role in reverse.

A surprisingly elegant retirement routine is *rebalance-aware withdrawals*: instead of selling a little of everything to generate spending cash, you preferentially sell from the asset class that is overweight relative to target. If stocks have had a strong run, you sell stocks to fund spending, which also trims risk. If stocks have fallen and bonds held up, you sell bonds or cash-like assets to fund spending, giving stocks time to recover.

This isn't a guarantee of success. It is a way to avoid unforced errors. It reduces the chance that you'll sell the exact asset that is currently depressed, which is the behavioral heart of sequence-of-returns damage.

In other words, rebalancing isn't just maintenance. It is also a *sourcing policy*.

### A realistic warning: taxes can turn rebalancing into self-sabotage

Rebalancing inside tax-advantaged accounts (like IRAs and 401(k)s) is relatively clean because trades don't trigger taxable events. In taxable brokerage accounts, rebalancing can be expensive if it triggers capital gains.

This is where good investors stop being ideological and start being practical. Rebalancing is not a religion. It is a tool. If the tool creates a large tax bill, you may need a more tax-aware approach:

- Use new contributions (or withdrawals) to rebalance whenever possible.
- Rebalance with dividends and interest flows.
- Prefer selling lots with losses or smaller gains when trimming a winner.
- Accept some drift if correcting it would cause outsized taxes.

The best rebalancing policy is the one that keeps you near your target allocation *net of real-world friction*. A pristine allocation that costs a fortune to maintain is not pristine; it is inefficient.

## Transaction costs and the ``picked off" problem: why rebalancing doesn't need to be predictable

There is a subtle modern critique worth knowing: if large pools of capital rebalance in predictable patterns (month-end, quarter-end), other market participants may anticipate that flow. You don't have to believe in shadowy conspiracies to understand the basic idea. If your behavior is obvious and scheduled, you may sometimes get worse pricing.

For individual investors, this is not a reason to panic. It is a reason to avoid needless precision. Rebalancing "on the last trading day of every quarter at 3:58 p.m." is not a virtue. It is an invitation to be a liquidity provider at inconvenient times.

A simple improvement: use a window rather than a point. If you rebalance annually, do it sometime in your birthday month. If you use bands, act when thresholds are breached, not when a calendar tells you to. If you have flexibility, place trades when markets are liquid and you can get reasonable spreads (especially for less liquid funds).

The goal is not to be clever. The goal is to be *unremarkable* in the market's eyes and disciplined in your own.

## Benchmarking: measuring performance without lying to yourself

Most investors benchmark themselves against the S&P 500, which is like benchmarking your health against an Olympic sprinter. It is not that the sprinter is bad. It is that the comparison is mismatched.

If you hold a diversified portfolio with bonds, international stocks, and perhaps real assets, your benchmark should reflect that mix. Otherwise you'll repeatedly draw the wrong conclusion:

- In bull markets, you'll feel "behind" because bonds dampen upside.

- In bear markets, you'll feel "smart" because bonds dampen losses.

Both feelings can be dangerous because they tempt you to abandon the allocation when it is doing exactly what it was designed to do.

A proper benchmark is simple: create a blended index that matches your target allocation. If you're 60/40, benchmark against a 60/40 blend. If you're 80/20, benchmark against 80/20. This isn't about bragging rights. It is about feedback. You want to know whether your portfolio is underperforming because of costs and implementation issues, or because you're comparing it to the wrong thing.

Rebalancing and benchmarking are cousins: rebalancing keeps risk aligned; benchmarking keeps expectations aligned.

### The behavioral hazard: rebalancing feels like ``selling winners'' and ``buying losers'' because it is

Rebalancing is easy to endorse in theory and hard to execute in practice because it asks you to do socially embarrassing things. When an asset class has been soaring, everyone at dinner is talking about it. Trimming it feels like leaving the party early. When an asset class is collapsing, buying it feels like walking into a building while people are running out.

This is where a written policy matters. A rebalancing plan is not merely a plan to trade. It is a plan to protect your future self from a very predictable emotional reaction: the urge to chase what's hot and abandon what hurts.

The best policy is specific enough that you can follow it without debate, but loose enough that you don't create unnecessary friction.

### A concrete rebalancing policy that works for real people

If you want something practical and robust, a hybrid approach tends to work well:

- Check allocations on a calendar (say, quarterly or semiannually).
- Only trade if drift exceeds bands (say, 5 percentage points for major asset classes).
- In taxable accounts, use contributions/withdrawals first; trade only when necessary.

This hybrid keeps you from overtrading during small noise while still preventing big drift from becoming a stealth risk bet. It also reduces the number of "should we rebalance?" conversations, which is an underrated benefit. Fewer conversations means fewer opportunities for fear to veto your process.

### Actionable takeaway

Write a one-paragraph rebalancing rule and put it where you will actually see it:

> My target allocation is ___ (e.g., 70/30 stocks/bonds with X% international). I will check my allocation ___ times per year. I will rebalance only when an asset class is more than ___ percentage points away from target (or ___

Rebalancing is not about being right about the market. It is about being consistent about yourself. The market will constantly try to change your portfolio's risk level without asking permission. A rebalancing discipline is how you keep ownership of the wheel.

Chapter 7

# The Automated Operating System

*Willpower is a depreciating asset; by Friday afternoon, it's usually gone. If your financial future relies on you manually logging in to transfer money or rebalance a portfolio, you are introducing a single point of failure: yourself. This chapter is about removing human emotion and inertia from the equation. We will build an automated infrastructure-a financial operating system-that executes your strategy flawlessly in the background, ensuring that compounding happens uninterrupted, regardless of whether you are busy, distracted, or asleep.*

## 7.1 The No-Willpower Workflow

On payday, most people perform a tiny magic trick: money arrives, and then it evaporates. Not in a dramatic "I bought a boat" way. In the quiet way. A few taps for delivery. A subscription that renewed because canceling would require remembering a password. A "we deserve this" dinner that costs exactly what your future self planned to invest. None of it feels like sabotage in the moment; it feels like life.

The problem isn't that you lack discipline. The problem is that you're asking the most distractible creature on Earth (a modern human with a phone) to repeatedly make a high-quality decision at the exact moment the rest of life is also demanding decisions. Every month you rebuild the same tower of good intentions—then wonder why a stiff breeze knocks it down.

A workflow is the opposite of intention. It's architecture.

The phrase *pay yourself first* is often delivered like a slogan, but it's really an engineering spec: the first claim on your income should be your long-term goals, not your short-term temptations. The trick is making that claim *automatic*, so the system behaves the same on a calm Tuesday and a chaotic Thursday. You want investing to feel less like a moral achievement and more like a utility bill—inevitable, boring, and quietly powerful.

A practical way to build that inevitability is a "daisy-chain" of accounts: money lands in one place, then gets swept along a pre-built path into its final destinations. Think of it like plumbing. You don't carry water around your house in a bucket. You install pipes once, then you live your life.

Here's the core idea: create one central *hub* account where income arrives (usually a checking account), and then build a set of *spokes*—automatic transfers and paycheck deductions—that route money out to specific "buckets" with jobs. Each bucket exists to eliminate a particular failure mode: running out of cash, skipping retirement contributions, procrastinating on investing, or accidentally spending what should have compounded for decades.

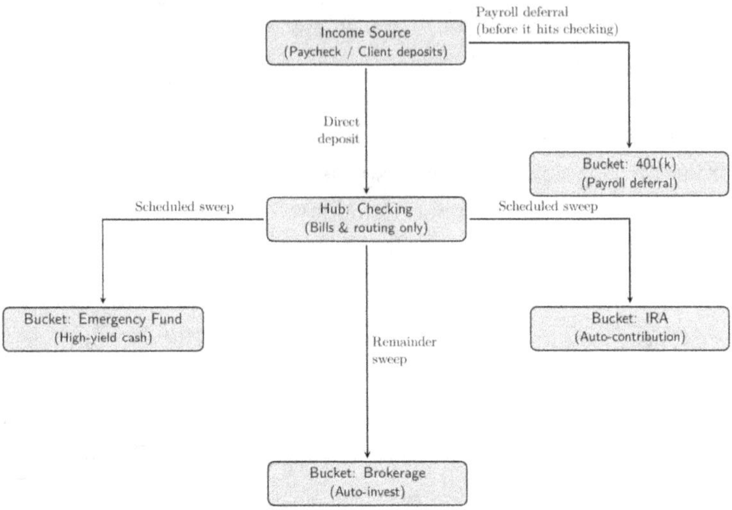


The power move is psychological as much as mathematical: *artificial scarcity*. You're deliberately making your spending account look smaller than your actual earning capacity. Not because you're trying to "trick" yourself in a childish way, but because your brain responds to what it sees. If your checking balance looks fat, you'll find "needs" to match it. If it looks lean-but-healthy, you'll adapt. Humans are remarkably good at living within the container provided-so choose the container.

A hub-and-spoke setup also reduces the number of moments where you can negotiate with yourself. Negotiation is where plans go to die. "I'll invest what's left at the end of the month" sounds responsible, but it's structurally flawed: the end of the month is when surprises have already happened. You're volunteering investing to be the shock absorber for everything else. The workflow flips that: investing happens early, and spending adapts around it.

*But what if the month really is tight?* Great—then your system reveals that quickly. A good financial system tells you the truth early enough to do something about it.

### The ordering rule: tax-advantaged first, then taxable

If you only remember one ordering principle, make it this: route dollars toward the accounts that give them the best "compounding conditions" first.

A *tax-advantaged account* is any account where the government gives you a deal—typically by letting investments grow tax-deferred, or by letting contributions reduce taxable income, or both. In the U.S., the usual lineup is employer retirement plans (like a 401(k)), IRAs, HSAs, and certain education accounts. A *taxable account* (like a regular brokerage account) has more flexibility, but fewer tax benefits.

Tax advantages matter because taxes are not just a cost; they're a drag on compounding. Compounding thrives when you leave it alone. Taxes force interruptions.

So the workflow usually runs like this:

- Capture the employer match (if available) because it's the rare return that doesn't require market optimism.
- Build a baseline emergency fund so you don't raid investments when life gets loud.
- Maximize or systematically fund retirement accounts.
- Then send overflow into taxable investing where you can scale contributions without ceiling constraints.

This isn't moralizing. It's just the cleanest way to turn each marginal dollar into the greatest expected long-term purchasing power, with the least friction.

### A story that ends the same way every month

Consider two versions of the same person-same salary, same rent, same taste for overpriced coffee.

*Version A* invests when they *get around to it*. Some months are great. Other months, work explodes, or a friend's wedding happens, or the car makes a sound that feels expensive. Investing becomes the thing that gets postponed because it *can* be postponed. They don't quit; they just delay. Years later, the tragedy isn't a single bad decision-it's the invisible pile of missed repetitions.

*Version B* gets paid and, within 24 hours, the money is already where it needs to be. They still buy coffee. They still have surprise expenses. But the default ending of each month is the same: contributions happened. The market did whatever the market did. Life did whatever life did. The long-term plan advanced anyway.

That's what you're building: a system where the "normal" outcome is the one your future self needs.

### Design the hub like a machine, not a lifestyle

Treat the hub checking account as a routing station, not a scorecard.

Two common mistakes:

1. *Using checking as a savings account.* This creates a false sense of wealth and invites casual spending.
2. *Letting checking become the place where decisions happen.* If you regularly "decide" what to transfer, your workflow is still powered by willpower.

A hub checking account should have three traits:

- *Predictable inflows.* Income should land there cleanly (or, even better, flow around it via payroll deferrals).
- *Known outflows.* Bills, rent/mortgage, and planned spending.
- *Automatic exports.* Scheduled sweeps that happen whether you're motivated or not.

If you want a simple mental model: checking is a conveyor belt. Money should not linger on a conveyor belt; it should move to where it's used or stored.

### Building the spokes: what each bucket is protecting you from

#### Emergency fund: protection from forced selling.

The emergency fund isn't there to earn a thrilling return. It's there to prevent a particular kind of financial self-harm: selling long-term investments at the worst possible time because the fridge died during a market drawdown.

In practice, automate the emergency fund like you'd automate a slow drip irrigation system. Small, consistent transfers until you hit a target, then the transfer stops (or gets redirected). You can set a calendar reminder once a year to review the target; you don't need to debate it weekly.

*A subtle psychological win:* an emergency fund reduces the emotional pressure you feel during volatility. When you know you can pay for real life without touching investments, market noise gets quieter.

#### 401(k): protection from the ``I'll do it later'' trap.

Payroll deferral is the cleanest form of automation because the money never becomes spendable in the first place. That's artificial scarcity at industrial strength.

If your plan offers a match, missing it is like declining part of your compensation. Even if you're skeptical of markets, even if you're paying down debt, even if you're "waiting for a better time"—the match is an immediate, contractually defined payoff for a behavior you want anyway: building long-duration capital.

A workflow note: treat your contribution rate like a dial you set, not a mood you express. Many people set it once (often too low) and then never touch it. Better: increase it on a schedule tied to raises. If you get a 4% raise, earmark 2% for your future self before your lifestyle learns to spend it.

### IRA: protection from job changes and plan limitations.

An IRA is often the "portable" bucket—the account that stays the same even if your employer changes, your payroll system changes, or your plan's investment menu is mediocre.

The automation challenge is that IRAs typically require you to push money from a bank account. That's exactly why the hub exists. Set a monthly transfer that occurs shortly after payday, and pair it with auto-invest inside the IRA so the cash doesn't accumulate in limbo.

If your income is irregular (commission, self-employment), the workflow still works—just change the rhythm. Use either:

- a fixed monthly transfer you can reliably sustain, plus occasional "bonus sweeps," or
- a percentage-based sweep from each inflow (some banks and cash management apps allow rules like this).

Irregular income doesn't prevent automation; it just demands a wider safety buffer in the hub so transfers don't bounce.

### Brokerage: protection from contribution ceilings.

At some point, the tax-advantaged buckets fill up-or your cash flow exceeds the allowed contribution limits. The brokerage account is where your system scales. It's also where behavior can get messy, because a brokerage account is one tap away from being a "market timing" playground.

So the rule is simple: make the brokerage account boring on purpose. Default to automatic, diversified purchases (often broad index funds or ETFs), and treat "extra" tinkering as a separate hobby budget—if you insist on having it at all.

### The investing toggle that matters more than your opinions

Most investors think their returns depend primarily on being right about the market. Over decades, your returns often depend more on whether your contributions happen relentlessly.

There's a reason automatic retirement plan enrollment changes outcomes: inertia is real. People stick with defaults—not because they're lazy, but because life is busy and the default feels like a recommendation. If you accept that about human nature, you stop trying to become a different species and start designing defaults worth sticking with.

A no-willpower workflow is the adult version of that insight: create defaults that keep your money moving toward compounding even when your attention is elsewhere.

### Avoiding the two classic automation face-plants

Automation is powerful, but it is not magical. It has predictable failure modes, and you can design around them.

#### Face-plant #1: overdrafts and cascade fees.

If you schedule aggressive sweeps out of checking without a buffer, one weird timing mismatch can trigger overdrafts, declined payments, and fees that feel like insult piled on injury.

A simple solution: maintain a *hub buffer*—a minimum checking balance that never gets swept. It's not an emergency fund; it's a shock absorber for timing. For many households, something like one month of core bills is plenty, but the right number is the one that makes the system stable under your real income timing.

If you're prone to variable expenses, build a second buffer: a "smoothing" savings account that receives irregular income and feeds checking on a schedule. This turns lumpy cash flow into predictable payroll for yourself.

### Face-plant #2: cash piling up uninvested.

Many people successfully automate transfers into investment accounts and then forget the final step: actually investing the cash. Months later, they discover they've built a respectable cash position… inside a brokerage account earning basically nothing.

The fix is unglamorous: ensure auto-invest is turned on, and confirm what it buys. If your platform can't auto-invest the specific ETF you want, you can often auto-invest a mutual fund equivalent, or schedule a recurring purchase.

This is one of those places where "pretty good" beats "perfect but manual." A slightly suboptimal fund you reliably buy for 10 years will usually beat an optimal fund you intended to buy.

## A concrete blueprint you can implement this weekend

If you want a workflow that feels like it runs itself, aim for a small number of moving parts that each do one job well.

| *Component* | *Default job description* |
|---|---|
| Income routing | Direct deposit to hub checking; payroll deferral to 401(k) happens upstream. |
| Hub checking | Pay bills; keep a hub buffer; execute scheduled sweeps; avoid accumulating "extra." |
| Emergency fund | Automated build to target; then maintenance-only transfers. |
| Retirement accounts | Automatic contributions; auto-invest enabled; simple, diversified holdings. |
| Brokerage account | Receives overflow; recurring purchases; minimal tinkering. |

None of this requires you to be a financial monk. It requires you to be a systems designer for about two hours, once.

### The hidden benefit: fewer identity-level decisions

There's a quiet psychological relief that shows up a few months after you automate: you stop repeatedly asking, "Am I the kind of person who invests?" You've already answered that question with wiring.

When investing depends on repeated self-affirmation, every month becomes an identity test. Identity tests are exhausting, and exhaustion breeds exceptions. A workflow turns investing into a background process. You can still be ambitious, still optimize, still learn—but your baseline wealth-building no longer depends on your mood.

### Actionable takeaway: set your ``48-hour rule" and one sweep

Pick one payday (or one day of the month if you're self-employed) and implement a single rule:

*Within 48 hours of income hitting the hub, money must leave the hub automatically.*

Concretely:

1. Set (or increase) your payroll deferral to retirement so it happens before money reaches checking.
2. Schedule one automatic transfer from checking to your IRA or brokerage for a date that reliably follows payday (e.g., payday + 1 business day).
3. Turn on auto-invest for that transfer so cash doesn't sit.
4. Add a hub buffer so the system can't trip on timing.

If you do only that, you've converted investing from a monthly decision into an always-on behavior—and compounding loves always-on behaviors.

## 7.2 Essential Plumbing and Hygiene

A well-built investing workflow can still fail for the dumbest reasons imaginable. Not because you panicked and sold the bottom. Not because you picked the "wrong" fund. Because somewhere in your account settings, a default dropdown quietly chose your future.

That's the theme here: *small toggles, large consequences.* Most of them feel administrative—like changing the batteries in a smoke detector. Which is exactly why they're so valuable. They don't ask you to become wiser or calmer. They just remove avoidable friction and prevent avoidable damage.

If automation is how money *arrives,* plumbing and hygiene are how money *behaves once it gets there.* The goal is not to create a perfect portfolio. The goal is to prevent your portfolio from accumulating invisible barnacles: cash that never gets invested, taxes that didn't need to be paid, records that become archaeology when it's time to file, rebalance, harvest losses, or move brokers.

### The three drags that never make headlines

Markets get blamed for a lot. But in long-duration investing, the most reliable performance killers are usually mundane:

- *Cash drag:* money sitting uninvested because dividends or contributions pile up as "buying power."
- *Tax drag:* paying taxes earlier than necessary, or paying more than necessary, because of how shares were sold or how distributions were handled.
- *Process drag:* time, confusion, and errors caused by missing documentation—often at the exact moment you least want a paperwork project.

These drags don't announce themselves with red banners. They show up as a life pattern: "My returns never seem to match what the market did." The market didn't cheat you. Your defaults did.

### Dividend reinvestment (DRIP): the tiny switch with compounding in its teeth

A dividend is a company sending some cash back to shareholders. In a brokerage account, that cash has to go somewhere. Your platform will usually offer two options:

- *Reinvest (DRIP):* automatically use the dividend to buy more shares.
- *Pay to cash:* deposit the dividend into the account's cash balance.

DRIP is boring. That's why it works.

When dividends go to cash, you've created a recurring decision point: "Should I invest this now?" Decision points are where money goes to die—because the cost of delaying feels like zero. In reality, delaying is the most expensive option because it steals time in the market. Not with drama, with drift.

Here's the part that surprises people: cash drag often isn't huge on any single dividend payment. It's huge in aggregate because it repeats. A small leak that happens four times a year for 25 years is not a small leak. It's an indoor pool.

There are, however, legitimate reasons *not* to DRIP:

- If you're intentionally redirecting dividends to rebalance (for example, using incoming cash to buy the underweight asset rather than automatically buying more of whatever paid the dividend).
- If you're living off portfolio income in retirement.

- If your broker's DRIP mechanics are poor (some brokers only DRIP whole shares; fractional shares are common now, but not universal).

But for most accumulators, DRIP is the default because it prevents a predictable failure mode: uninvested cash that quietly grows in proportion to your success.

**A practical hybrid that feels "adult."** If you like the *idea* of being deliberate—without relying on willpower—use this rule:

> *DRIP inside retirement accounts. In taxable accounts, either DRIP everything or DRIP nothing—then schedule a recurring buy that invests all cash monthly.*

Why the bluntness? Because "I'll manually invest dividends when I notice them" is the financial equivalent of "I'll floss when I remember." A few people do it. Most people buy more mouthwash and call it self-care.

### Cost basis and tax lots: the stealth control panel for your future tax bill

If dividends are the small switch, tax lots are the hidden control room.

When you buy shares over time—especially with automatic investing—you don't own "a position." You own *layers* of that position purchased at different prices on different dates. Each layer is a *tax lot*. The price you paid is your *cost basis*. When you sell, the difference between the sale price and the cost basis determines your capital gain (or loss), and that determines how much tax you owe.

The brokerage has to choose *which shares you sold*. That's not philosophy; it's math. And the default choice may not be the best one for you.

Most brokerages let you choose a cost basis method for taxable accounts. Common options include:

- *FIFO (First In, First Out):* sells the oldest shares first.
- *LIFO (Last In, First Out):* sells the newest shares first.
- *Average cost:* uses the average purchase price (common for mutual funds; can be restrictive later).
- *Specific ID (Specific Identification):* you choose the exact lots you're selling.

If you want maximum flexibility, the setting to look for is usually *Specific ID.*

Specific ID is less "I love paperwork" and more "I want options." It allows you to choose the lots with:

- the *highest cost basis* to minimize taxable gains when you need cash,
- the *lowest cost basis* if you're doing charitable giving of appreciated shares,
- or the lots with *losses* if you're harvesting tax losses.

The point isn't to become a tax wizard. The point is to avoid locking yourself into a blunt instrument when you could have had a scalpel.

**A story: two investors sell the same dollar amount and get two different tax bills.** Imagine two people each sell $20,000 of the same index fund in a taxable account to make a down payment.

Investor 1 uses FIFO by default. Their oldest shares were bought cheap years ago. They realize a large gain. Taxes take a bigger bite than expected, so they sell a little more to cover the tax, realizing more gain, paying more tax. It's not catastrophic. It's just annoying and expensive in the specific way that makes you distrust the whole system.

Investor 2 uses Specific ID. They sell higher-basis lots accumulated more recently (still long-term, but purchased at higher prices). They realize a smaller gain for the same $20,000 proceeds. Same market. Same fund. Different tax result—because one person controlled which shares left the building.

That's "plumbing." Nobody brags about it at dinner. But it can save real money and preserve more shares to keep compounding.

**Two cautions that keep you out of trouble.**

- *Specific ID only works if you actually select lots at sale time.* Some brokers let you set a default method but still require lot selection on each sale. Make sure you know the workflow before you need it quickly.
- *Don't use Average Cost casually.* Once you elect it for certain holdings (often mutual funds), switching later can be limited. It's not "bad," it's just less flexible.

### The wash sale boomerang: when a tax loss isn't really a loss

Tax-loss harvesting is the practice of selling an investment at a loss to offset capital gains (and potentially a limited amount of ordinary income), then reinvesting in a similar—but not "substantially identical"—investment to maintain market exposure.

The boomerang is the *wash sale rule*. If you sell a security at a loss and buy the same or substantially identical security within a window around that sale, the loss can be disallowed and added to the cost basis of the replacement shares.

Here's where hygiene matters: automatic investing can accidentally trigger wash sales.

Example: you sell an ETF at a loss on December 10 to harvest the loss. Meanwhile, your account is set to auto-buy that same ETF every two weeks, and it purchases it on December 15. Congratulations: you've created a wash sale problem with your own robot.

This isn't a reason to fear automation. It's a reason to keep a short checklist for "tax actions":

- Pause auto-invest into that ticker before harvesting.
- Consider buying a similar (but not identical) fund as the replacement.
- Resume automation after the window has cleared.

The emotional benefit of knowing this rule is underrated. It lets you act deliberately without feeling like taxes are a trapdoor waiting for you.

### Rebalancing without drama: aim your new money like a thermostat

Rebalancing means bringing your portfolio back toward a target mix—say, 80% stocks and 20% bonds—after market movements push it off course. It's risk management disguised as housekeeping.

The hygiene angle is simple: the cleanest rebalancing is often done with *new contributions*, not sales.

Why? Selling in a taxable account can trigger capital gains taxes. Redirecting your next contributions toward the underweight asset can move your allocation back toward target without creating a tax bill. It's like steering a boat by adjusting the engine instead of throwing cargo overboard.

This plays beautifully with an automated workflow. You can set your retirement accounts to a target allocation and let contributions flow into those funds. In taxable accounts, you can:

- contribute to a single "all-in-one" fund if appropriate (simplifies rebalancing),
- or alternate contributions between funds based on which is underweight,
- or maintain a small spreadsheet or note that tells you what the next month's auto-buy should be.

The goal isn't to rebalance constantly. The goal is to avoid waking up five years later with a portfolio that quietly mutated into something riskier than you intended—usually because stocks did well and you never trimmed or counterweighted.

### Cash management: the difference between ``push" and ``pull" money

Not all automation is created equal.

A *push payment* is when you instruct your bank to send money out (like bill pay or an ACH transfer you initiate). A *pull payment* is when you authorize a company to take money from your account (like many subscription services or some brokerage funding arrangements).

From a hygiene standpoint, push is often easier to control. Pull can be convenient, but it can also create overdraft surprises if timing or amounts vary and your hub buffer is too thin.

This matters because brokerages increasingly offer "cash management" features that blur the line between bank and brokerage. The convenience is real—but so is the operational risk of having too many entities with permission to pull cash from the same account.

A clean setup tends to look like this:

- Bills and obligations are paid in predictable ways from the hub.
- Investing contributions are pushed out on a schedule you control.
- Any pull-based authorizations are limited, reviewed, and removed when no longer needed.

Hygiene isn't paranoia. It's simply keeping the number of ways money can leave your accounts smaller than the number of ways it can enter.

### Statements, confirmations, and the one folder your future self will worship

The most expensive financial mistakes are sometimes administrative, not analytical.

Brokerages will track cost basis and generate tax forms. Employers will issue W-2s. Fund companies will report distributions. Most of the time, it works.

And then you transfer brokers. Or you inherit an account. Or you do a backdoor IRA contribution. Or your CPA asks for a confirmation statement from a transaction two years ago. Or the IRS sends a letter because their data doesn't match your reporting. Or you need to prove your cost basis for shares acquired long ago.

At that moment, your memory becomes a liar, and your email becomes a landfill.

A simple hygiene system:

- Create a folder structure (cloud + local backup) with year subfolders.
- Save monthly statements and all trade confirmations for taxable accounts.
- Save year-end tax forms (1099, 5498, 1098, etc.) and contribution confirmations.
- Keep a single "Account Map" document listing: institutions, account types, last four digits, beneficiaries, and where the records live.

This is not about being organized for its own sake. It's about preserving optionality. When opportunities or problems appear, organized investors move faster—and speed is a form of return.

### Beneficiaries: the checkbox that can override your will

A beneficiary designation is the person (or trust) that receives an account when you die. It sounds grim. It's actually love—expressed through paperwork.

Here's the uncomfortable truth: beneficiary forms can override your will. You can have a beautifully drafted estate plan, and one stale beneficiary designation can send assets somewhere you didn't intend, creating family conflict with an interest rate.

Retirement accounts and life insurance are especially beneficiary-driven. Hygiene means:

- Set beneficiaries when the account is opened.
- Review them after major life events (marriage, divorce, children, deaths).
- Keep a copy of the confirmation page or statement showing the designation.

If you want a rule that's hard to ignore: *if you would update your emergency contact, update your beneficiaries.*

### ``Set and forget" does not mean ``set and never verify"

Automation deserves one kind of ongoing attention: periodic audits.

Not frequent tinkering. Audits.

Once or twice a year, run a simple inspection:

- Are contributions still happening on schedule?
- Is cash accumulating uninvested?
- Are dividends being handled the way you intend (DRIP vs cash)?
- Is cost basis method set the way you want for taxable accounts?

- Did any accounts change features, migrate platforms, or reset settings after an update?
- Are beneficiaries still correct?

This is how you keep a machine a machine. Even the best systems degrade if you never look at them. The point is to look rarely, but effectively.

### Actionable takeaway: do the 30-minute ``account hygiene sweep"

Open your primary brokerage (and your retirement plan portal if you have one) and change or confirm these settings today:

1. *Dividend handling:* choose DRIP or a deliberate alternative that still auto-invests cash on a schedule.
2. *Cost basis method (taxable accounts):* set to *Specific ID* if available and appropriate for you.
3. *Auto-invest:* confirm recurring purchases are active and buying the intended funds (not just accumulating cash).
4. *Beneficiaries:* verify they exist and reflect current reality; save a confirmation.
5. *Records:* download the latest statement and create the folder where all future statements will go.

If this feels unglamorous, good. Unsexy habits are often the ones that survive for decades—and decades are where compounding does its best work.

## 7.3 Firewalling Your Wealth

Market risk gets the spotlight. Platform risk gets the fine print.

Platform risk is the chance that your money is *fine in theory* but inaccessible, missing, frozen, delayed, misdirected, or stolen in practice—because the broker is down, your credentials are compromised, a

SIM swap lets someone intercept your login codes, a payment link you clicked was a counterfeit, or a real institution fails and the cleanup takes months. None of these scenarios require you to be reckless. They just require you to be human in a world where crime is automated and outages are normal.

Compounding is fragile in a specific way: you can't compound what you can't keep. A 30% drawdown is painful but historically survivable for diversified investors with time. A total loss from fraud is a different species of event. It doesn't feel like "investing." It feels like having the floor removed.

A good firewall doesn't promise invincibility. It creates layers—so one mistake or one failure doesn't become a full account wipe. That mindset is borrowed from engineering and cybersecurity: *defense in depth*. You assume something will break. You design so the break is contained.

**Defense in depth (security checklist begins here)**

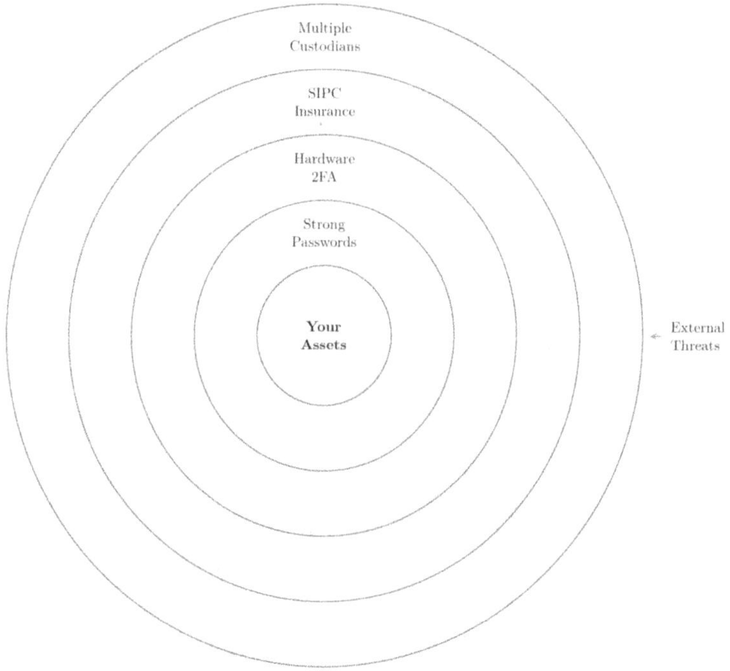


The diagram is intentionally simple because the philosophy is simple: you want multiple independent "no" votes standing between an attacker (or a failure) and your assets. If you only have one gate—one password, one device, one brokerage, one email account—then every problem is a *single point of failure*. Single points of failure are how small mistakes become permanent.

The firewall breaks into two categories: controls that protect *access* (so nobody can move your money), and controls that protect *custody* (so even if a platform fails, your assets aren't stranded beyond recovery). Both matter. One keeps thieves out. The other keeps you functional during chaos.

### Layer 1: strong passwords (the boring lock that actually works)

A weak password isn't just guessable; it's *reusable.* Reuse is the real killer. Data breaches leak billions of credentials. Attackers don't "hack" your brokerage first. They try your email and password everywhere until something opens—like a burglar walking down a street trying doorknobs.

A strong password strategy is less about genius and more about consistency:

- Use a password manager so every important account gets a unique, long password.
- Protect the password manager itself like it's the crown jewels, because it is.
- Treat your email account as your master key; if someone controls your email, password resets become a vending machine.

There's also a counterintuitive point that good standards now recognize: forcing arbitrary password changes every 90 days can make people *less* secure, because it trains predictable patterns ("Password-Summer2026!"). Better: long, unique passwords; change them when there's reason (breach, compromise, risk event), not as a ritual.

If you want a mental model, think of password strength like a seatbelt. You don't put it on because you plan to crash. You put it on because you plan to be alive in a world where crashes happen.

### Layer 2: hardware 2FA (the deadbolt thieves hate)

Two-factor authentication (2FA) means you need something *besides* a password to log in—often a code generated by an app or a physical security key. This matters because passwords are copyable. Your phone and hardware keys are harder to copy at scale.

The highest-friction option for an attacker—and therefore the best for you—is usually a *hardware security key* (often called a "hardware key" or "security key"). It's a small device that plugs into your computer or taps your phone. Without the physical key, the login can't complete.

Why bother, when an SMS code is "good enough"? Because SMS is not a security system; it's a convenience system. Phone numbers can be hijacked through SIM swap fraud or carrier account compromise. If your brokerage relies on SMS codes, a criminal can aim at your phone carrier instead of your broker.

A practical setup that doesn't require paranoia:

- Use a hardware key for your email account and your brokerage if supported.
- Keep a second backup hardware key stored separately (not on the same keychain), because loss happens.
- If hardware keys aren't supported, use an authenticator app rather than SMS whenever possible.

This is one of the rare upgrades where the inconvenience is a feature. You're adding a physical step that makes remote theft dramatically harder.

### The "automation paradox": your system gets stronger and your attack surface gets bigger

Automation reduces the number of decisions you have to make. It can also increase the number of *connections* between accounts—bank links, brokerage links, payroll portals, budgeting apps, aggregation dashboards.

Connections are useful. Connections are also doors.

The firewall approach isn't "never connect anything." It's "connect with intent":

- Link accounts where it removes repeated manual handling (high value).
- Avoid granting broad permissions to low-trust third-party apps (low value).
- Periodically audit "linked accounts" and revoke anything you no longer use.

In other words: you want a streamlined machine, not an over-networked science project.

### Layer 3: know what's insured (and what isn't)

Insurance is part of a firewall, but only if you understand what it covers. Otherwise, it's a bedtime story you tell yourself so you can stop thinking about risk.

Two acronyms get confused constantly:

**FDIC: bank deposits.** FDIC insurance applies to *deposit accounts* at insured banks—checking, savings, CDs—up to coverage limits and rules. It's designed for bank failures. It is not designed to cover investment losses, and it doesn't apply to stocks and bonds you hold at a brokerage.

**SIPC: brokerage custody shortfalls.** SIPC protection applies to *missing* cash and securities at a SIPC-member brokerage, up to limits (including a cash sublimit). The key word is missing. SIPC is about restoring what should be there if the broker fails and assets are not properly segregated or are otherwise unavailable. SIPC does *not* protect you from market declines. It does *not* make a bad investment good. It's not investment insurance; it's custody insurance.

This distinction matters because it changes how you behave:

- If you're holding cash you need soon, where that cash sits (bank vs brokerage cash sweep vs money market fund) determines what protections may apply.
- If you're holding long-term securities at a brokerage, SIPC is a backstop for operational failure, not a reason to ignore fraud prevention.

The practical emotional benefit of getting this right is huge. Clear understanding reduces panic during headlines. When a financial institution has a bad news cycle, you want to know whether you're facing a *market* issue, a *bank* issue, a *brokerage custody* issue, or simply an *outage* and a dopamine-saturated internet.

### Layer 4: multiple custodians (institutional diversification)

Diversification usually means "own lots of companies." Institutional diversification means "don't have your entire financial life depend on one login."

Even very good institutions can have outages. Even honest institutions can get hit by cyber incidents. Even well-regulated institutions can fail. And even if everything is ultimately fine, a temporary freeze can be a problem if you need liquidity to pay taxes, fund a home purchase, or simply make payroll for your own business.

Splitting assets across multiple custodians is not about expecting disaster. It's about ensuring continuity.

A simple, sane version looks like:

- One primary brokerage for most investing.
- A second brokerage (or a bank + brokerage combination) holding a smaller "continuity sleeve": enough cash or liquid securities to operate for a while if the primary platform is inaccessible.

High-net-worth investors often take this further to manage coverage limits and to reduce concentration in any single institution's operational risk. But you don't need to be ultra-wealthy to benefit from not being one password away from paralysis.

**A realistic scenario: the week your broker is "temporarily unavailable."** Imagine you're trying to wire earnest money for a home purchase. Your broker is the source of the down payment cash. That morning, the platform is down—maybe from an outage, maybe from a fraud lock that triggered because you traveled, maybe from a "routine security review."

It may resolve in hours. It may resolve in days. The bank doesn't care; the deadline is still the deadline.

Multiple custodians aren't an investing flex. They're a life logistics tool.

### Identity theft prevention: the credit freeze that does more than you think

Many financial frauds don't start with stealing your existing accounts. They start with opening new ones in your name.

A *credit freeze* is one of the strongest preventive controls available to ordinary people. It restricts access to your credit report so new lenders can't easily extend credit in your name. It's typically free, doesn't affect your credit score, and stays in place until you lift it (temporarily or permanently), but you must place it with each major credit bureau.

The reason this belongs in a wealth firewall is simple: identity theft can create a mess that bleeds into your investing life—new accounts, mail redirection, address changes, bank account takeovers, and a flurry of "verification" events that train you to click the wrong link at the wrong time.

A credit freeze is like putting your identity behind glass. You can still access it. Random strangers can't.

### The operational checklist that prevents "stolen in silence"

Fraud often wins by being quiet. The goal is to move money without triggering your attention. So your firewall needs alarms.

Useful alarms aren't complicated:

- Turn on transaction alerts for withdrawals, new payees, profile changes, and new device logins.
- Enable login notifications for your email and brokerage.
- Review account activity regularly enough that "something weird" stands out (monthly is often plenty for long-term investors).

The important nuance: alerts should go somewhere an attacker can't easily control. If your email is compromised, email-only alerts can be neutralized. Some platforms allow SMS or app notifications; use redundancy, but avoid relying solely on SMS for authentication.

### Make your "money-moving" process inconvenient on purpose

A lot of personal finance advice chases convenience. Security sometimes requires the opposite: intentional inconvenience, placed only where it matters.

A powerful pattern:

- Keep investing automated and smooth (good frictionlessness).
- Make withdrawals and transfers *high-friction* (good friction).

This means:

- Use hardware 2FA for logins.
- Use verbal passphrases where available.
- Limit wire capability or require extra verification.
- Avoid storing bank-linked withdrawal instructions everywhere "just in case."

You're building a system where it's easy for *you* to fund your future, and hard for *anyone* to evacuate it.

### The emotional skill: never authenticate an inbound request

Most fraud now is social engineering: convincing you to authorize the theft yourself. The criminal's best tool is urgency—*act now or lose access*—and legitimacy cosplay—logos, spoofed numbers, fake "support" chats.

One rule is worth memorizing because it's so clean:

> *Never authenticate an inbound request.*

If you receive a call, text, or email claiming to be your bank or broker, do not click. Do not log in through the provided link. Do not read back codes. Instead, hang up and contact the institution using a known-good method: the number on the back of your card, the official app, or a bookmarked site you typed yourself.

This habit feels "rude" until the day it saves you.

### Actionable takeaway: a one-hour wealth firewall upgrade

Set a timer for one hour and implement the following, in order of impact:

1. *Secure your email first:* unique long password + 2FA (ideally hardware key). Email is the reset button for everything.

2. *Secure your brokerage login:* unique password + 2FA (prefer hardware key or authenticator app; avoid SMS when possible).
3. *Turn on critical alerts:* withdrawals, new payees, profile changes, and new device logins.
4. *Place a credit freeze:* do it at each major bureau so new-account fraud becomes harder.
5. *Create continuity:* open (or identify) a second custodian and keep a modest continuity sleeve so one outage can't halt your financial life.

Do that, and you've built something rare: a compounding machine that doesn't just grow—it *stays yours*.